I0119030

R. Price Williams

On the Increase of Population in England and Wales

Read before the Statistical Society, June 15, 1880

R. Price Williams

On the Increase of Population in England and Wales
Read before the Statistical Society, June 15, 1880

ISBN/EAN: 9783337325749

Printed in Europe, USA, Canada, Australia, Japan

Cover: Foto ©Suzi / pixelio.de

More available books at **www.hansebooks.com**

THE LONDON LIBRARY,

12, ST. JAMES'S SQUARE, S.W.

Patron.

HIS ROYAL HIGHNESS THE PRINCE OF WALES.

President.

THOMAS CARLYLE, Esq.

Vice-Presidents.

THE DEAN OF WESTMINSTER.	EDWARD H. BUNBURY, Esq.
RT. HON. W. E. GLADSTONE, M.P.	JAMES SPEDDING, Esq.

Trustees.

LORD HOUGHTON.	EARL OF CARNARVON.	EARL OF ROSEBERY.

Committee.

Sir James Alderson.	Sydney Gedge, Esq.	Jas. Cotter Morison, Esq.
F. W. Burton, Esq.	Professor Gladstone.	The Earl of Morley.
Rev. Canon Cheetham.	F. Harrison, Esq.	Dr. Munk.
J. C. Conybeare, Esq.	Dr. J. J. Jusserand.	Rev. Mark Pattison.
W. J. Courthope, Esq.	C. M. Kennedy, Esq.	F. Pollock, Esq.
Sir Frederick Elliot.	Rev. Dr. Stanley Leathes.	Rev. Dr. Reynolds.
Rev. E. E. Estcourt.	W. Watkiss Lloyd, Esq.	Herbert Spencer, Esq.
H. W. Freeland, Esq.	H. Maxwell Lyte, Esq.	Leslie Stephen, Esq.

The Library (established in 1841) contains 90,000 volumes of Ancient and Modern Literature, in various Languages : Subscriptions, £3 a-year, or £2 with Entrance Fee of £6. Life Membership, £26. Fifteen volumes are allowed to Country and Ten to Town Members. Reading-rooms open from ten to half-past six. Catalogue, New Edition, 1875 (1062 pp.), price 16s. ; to Members, 12s. Prospectuses on application.

ROBERT HARRISON, *Secretary and Librarian.*

---

*Reprinted from the Journal of the Statistical Society for 1851, Price 1s.,
with a* PREFACE *and* NOTES.

STATISTICS

OF THE

FARM SCHOOL SYSTEM

OF THE

CONTINENT,

AND OF ITS APPLICABILITY TO THE

PREVENTIVE AND REFORMATORY EDUCATION

OF

PAUPER AND CRIMINAL CHILDREN IN ENGLAND.

BY THE LATE JOSEPH FLETCHER, Esq.,

BARRISTER-AT-LAW, HONORARY SECRETARY.

LONDON: E. STANFORD, 55, CHARING CROSS, S.W

INCREASE OF POPULATION

IN

ENGLAND AND WALES.

From the JOURNAL OF THE STATISTICAL SOCIETY.
SEPTEMBER, 1880.

On the INCREASE of POPULATION in ENGLAND and WALES.
By R. PRICE WILLIAMS, M. INST. C.E.

[Read before the Statistical Society, 15th June, 1880.]

CONTENTS :

IN order to realise the enormous increase in the population of England and Wales during the present century, it is only necessary to compare it with that of the preceding one.

In the absence of any census returns, the amount of the population prior to 1801, can only be approximately arrived at, from the registers of births and deaths, and the poll and hearth tax returns : the following results obtained in this way, under what are known as the " Population Acts," afford the means of approximately determining the amount of the population in each decade :—

Population of England and Wales, 1700-1801.

Year.	Population.	Increase.	Decrease.	Percentage per Decade.
	No.	No.	No.	Per cnt.
1700	5,475,000	—	—	
				Decrease.
'10	5,240,000	—	235,000	− 4·92
				Increase.
'20	5,565,000	325,000	—	6·20
'30	5,796,000	231,000	—	4·15
'40	6,064,000	268,000	—	4·62
'50	6,467,000	403,000	—	6·64
'60	6,736,000	269,000	—	4·16
'70	7,428,000	692,000	—	10·27
'80	7,953,000	525,000	—	7·07
'90	8,675,000	722,000	—	9·08
1801	8,892,536	217,536	—	2·51
		3,652,536	235,000	+ 54·70
		− 235,000		− 4·92
		3,417,536	—	+ 49·78
Average decennial rate of increase				+ 4·98
Average taking initial and terminal				4·97

Although the results given in this table disprove the contention of Dr. Price, that the population had been continually declining in England from the period of the Revolution until 1777, they show very clearly that there was a considerable decrease in the first decade, and that the increase during the first half of the century was very slight; this is further confirmed by the small increase in the value of the exports in the same period, as shown in the following table :—

Year.	Exports.	Increase.	Rate of Increase per Decade.	Imports.	Increase.	Rate of Increase per Decade.
	£	£	Per cent.	£	£	Per cent.
1700	6,097,120	—	—	4,753,777	—	—
'50	10,130,990	4,033,871	10·69	7,289,582	2,435,805	8·93
'75	16,326,363	6,195,372	21·03	14,815,855	7,526,273	32·80
1800	38,120,120	22,793,767	40·38	30,570,605	15,745,750	33·61

The total increase in the value of the exports during the first fifty years it will be seen only amounted to 4,033,871*l.*, or to an average increase of 10·69 per cent. in each decade. The rate of increase, however, in the next twenty-five years was much more rapid, while during the last twenty-five years it amounted to as much as 40 per cent. per decade.

It was not in fact until the first decade of the second half of the last century, which witnessed the conclusion of the Seven Years' War, the invention and introduction into common use of the spinning jenny by Hargreaves, and the invention by Watt of his earliest form of steam engine, that any indications are afforded of that rapid increase of population which has so markedly characterised the present century; the increase in the decade 1760-70, however, amounted to as much as 10·27 per cent. In the diminished rate of increase of the next decade viz., 7·07 per cent., may be traced the effects of the war with America which begun in 1773, and that which shortly followed with France.

It is however worthy of note, that it was about the middle of this particular decade that Watt's steam engine, which has indirectly had so large a share in promoting the increase of the population of this country, began to be generally used on a large scale in the manufacturing districts.* The effects of the general adoption of steam power in manufactories, and of improvements in machinery, are clearly indicated in the higher rate of increase of the population (9·08 per cent.) which obtained in the next decade (1780-90); and concurrently with all this, it should be observed that the country was beginning to recover from the effects of the great stagnation of trade due to the American War.

* 1775.— Watt's engines erected upon a large scale in manufactories, and his patent renewed by Act of Parliament.

The increase during the last ten years of the century, marked as it was, as a period of war abroad, bad harvests* and troubles at home was—with the exception of the actual decrease in the first décade—the smallest which had occurred in any decade during the century.

The total increase in the population during the whole century only amounted to 3,417,536, giving an average of 4·97 per cent. per decade, an amount of increase which, as will be seen from the following table, it took little more than the two first decades of the present century to equal:—

Population of England and Wales, 1801-1871.

Period.	Population.	Amount of Increase.	Percentage of Increase.
			Per cnt.
1801	8,892,536	—	—
'11	10,164,256	1,271,720	14·30
'21	12,000,236	1,835,980	18·06
'31	13,896,797	1,896,561	15·81
'41	15,909,132	2,012,335	14·48
'51	17,927,609	2,018,477	12·69
'61	20,066,224	2,138,615	11·93
'71	22,712,266	2,646,042	13·19
	—	13,819,730	—

During the first decade of the present century (1801-11) the population of England and Wales increased as much as 14·30 per cent., while in the next decade (1811-21) it reached the maximum attained in this century, viz., 18·06 per cent.: as from that period down to the census of 1861, the rate of increase of the population has continuously diminished, the decrement during the several decades being as follows:—

Decade.	Rate of Increase.	Decrement in Rate of Increase Expressed as a Percentage on the Rate of Increase.	Increment in Rate of Increase.
	Per cnt.	Per cnt.	Per cnt.
1801–11	14·30	—	—
'11–21	18·06 ⎫		
'21–31	15·81 ⎪	− 12·46	—
'31–41	14·48 ⎬	− 8·41	—
'41–51	12·69 ⎪	− 12·36	—
'51–61	11·93 ⎪	− 5·99	—
'61–71	13·19 ⎭	—	+ 10·56
		− 39·22	+ 10·56
		+ 10·56	
		5) − 28·66	
Average decrement.................		− 5·73	
Average decrement for the five decades taking the initial and terminals		= 6·09	

* 1795 and 1797.

A remarkable increment in the rate of increase occurred, however, during the last decade (1861-71), to which further reference will be made.

In order to ascertain the respective rates of increase or decrease of the town and rural population, together with the decrements in their rates of increase, the writer has for some years past been engaged on the analysis of the census returns contained in the accompanying tables, viz. :—

> *Tables A^1 to A^{52}.**—Population, and its rates of increase or decrease per cent. of each county in England and Wales for each decade from 1801 to 1871, subdivided as follows :—
>
>> 1st.—Large towns, containing a population of 20,000 inhabitants and upwards.
>>
>> 2nd.—Small towns, containing a population from 2,000 to 20,000 inhabitants.
>>
>> 3rd.—Rural districts, including small towns or places with less than 2,000 inhabitants.
>
> *Table B.†*—Showing the aggregate population of the large towns of 20,000 inhabitants and upwards in each county in England and Wales, and the rates of increase or decrease per cent. for each decade 1801-71. (Summarised from Tables A^1 to A^{52}.)
>
> *Table C.†*—Showing the aggregate population of the small towns containing from 2,000 and under 20,000 inhabitants in each county in England and Wales, and the rates of increase or decrease per cent. for each decade 1801-71. (Summarised from Tables A^1 to A^{52}.)
>
> *Table D.†*—Showing the aggregate population of large and small towns combined, in each county in England and Wales, and the rates of increase or decrease per cent. for each decade 1801-71. (Summarised from Tables A^1 to A^{52}.)
>
> *Table E.†*—Showing the aggregate population of the rural districts, including small towns or places with less than 2,000 inhabitants in each county in England and Wales, and the rates of increase or decrease for each decade 1801-71. (Summarised from Tables A^1 to A^{52}.)
>
> *Table F.†*—Showing the total population and the rates of increase or decrease per cent. of each county in England and Wales for each decade 1801-71. (Summarised from Tables A^1 to A^{52}.)
>
> *Table G.†*—Showing the population and rates of increase or decrease per cent. of each town of 20,000 inhabitants and

---

* These tables are not printed.   † See Appendix.

upwards, in England and Wales, for each decade 1801-71. (Summarised from Tables A[1] to A[52].)

Population and Rates of Increase in the Principal Towns.

Attention has already been drawn to the fact that the maximum rate of increase of the entire population of England and Wales (18·06 per cent.) occurred in the decade 1811-21. In the case of

Summary of

Decade.	Large Towns over 20,000.				Small Towns over 2,000 and under 20,000.				Total
	Population.	Rate of Increase per Cent.	Decrement in Rate of Increase.	Increment in Rate of Increase.	Population.	Rate of Increase per Cent.	Decrement in Rate of Increase.	Increment in Rate of Increase.	Population.
			Per cent.	Per cent.			Per cent.	Per cent.	
1801.........	2,404,153	19·71			1,211,092	13·10			3,615,245
'11.........	2,878,039	24·46	—		1,369,757	*19·00			4,247,796
'21.........	3,582,029	*26·19	—		1,630,046	14·98	−21·16		5,212,075
'31.........	4,520,055	23·28	−11·11		1,874,112	12·46	−16·82		6,394,167
'41.........	5,572,175	23·56	—	+1·20	2,107,562	10·50	−15·74		7,679,737
'51.........	6,885,001	19·36	−17·83		2,328,941	7·31	−30·38		9,213,942
'61.........	8,218,269	*19·25	− 0·57		2,499,051	*11·07		+51·44	10,717,260
'71.........	9,800,887				2,775,739				12,576,626
			−29·51 +1·20	+1·20			−84·10 +51·44	+51·44	
			4)−28·31				5)−32·66		
Average decrement			− 7·08	—	—	—	− 6·53	—	—
*Average decrement, initial and terminal			− 7·40	—	—	—	−10·24	—	—

The rates of increase of nearly all the most populous towns such as Liverpool, Manchester, Leeds, and Birmingham, attained their maximum in the same decade as in the case of the aggregate town population, viz., 1821-31, the rate of increase of the population in the aggregate population of the larger towns of over 20,000 inhabitants, it will be found that the maximum rate of increase, viz., 26·19 per cent., occurred a decade later (1821-31): the rate of increase from that time having continually diminished, the decremental rate has, however, varied considerably, being as much as 11·11 per cent. in the following decade (1831-41), while in the next there was even a slight increment in the rate of increase of 1·20 per cent. as shown in the following table :—

Tables B *to* E.

Town Population.			Rural Districts and Small Towns under 2,000.				Total Population of England and Wales.			
Rate of Increase per Cent.	Decrement in Rate of Increase.	Increment in Rate of Increase.	Population.	Rate of Increase per Cent.	Decrement in Rate of Increase.	Increment in Rate of Increase.	Population.	Rate of Increase per Cent.	Decrement in Rate of Increase.	Increment in Rate of Increase.
	Per cnt.	Per cnt.	5,277,291			Per cnt.	Per cnt.	8,892,536		Per cnt.	Per cnt.
17·50	—			12·11	—			14·30		
22·70	} Nil		5,916,460	*14·74			10,164,256	*18·06	} −12·46	
*22·69			6,788,661		−28·63		12,000,236	15·81		
20·10	−11·42		7,502,630	10·52	− 7·89		13,896,797	14·48	} − 8·41	
19·98	− 0·60		8,329,395	9·69	−39·32		15,909,132	12·69	} −12·36	
16·32	−18·32		8,713,067	5·88		} +23·98	17,927,609	11·93	} − 5·99	
*17·35	—	} +6·31	9,348,964	7·29		} +15·36	20,066,224	*13·19	—	} +10·56
			10,135,640	*8·41			22,712,266			
	−30·34 +6·31	+6·31			−75·84 +39·34	+39·34			−39·22 +10·56	+10·56
	4)−24·03				5)−36·50				5)−28·66	
—	− 6·00	—	—	—	− 7·30	—	—	—	− 5·73	—
—	− 6·49	—	—	—	−10·62	—	—	—	− 6·09	—

those large towns, constituting about 75 per cent. of the entire population of the large towns, practically governing the period of the maximum rate of increase, as will be seen from a reference to the following table :—

Population and Rates of Increase of Population in the Principal

	1801.	1811.		1821.		1831.	
	Population.	Population.	Rate of Increase	Population.	Rate of Increase.	Population.	Rate of Increase.
London	958,863	1,138,815	18·77	1,378,947	21·09	1,654,994	20·02
Liverpool	82,295	104,104	26·50	138,354	32·90	201,751	45·82
Manchester	76,788	91,130	18·68	129,035	41·60	187,022	44·93
Birmingham	70,670	82,753	17·10	101,722	22·92	143,986	41·55
Leeds	53,162	62,534	17·63	83,796	34·00	123,393	47·25
Sheffield	45,755	53,231	16·34	65,275	22·63	91,692	40·47
Bristol	61,153	71,433	16·81	85,108	19·14	104,408	22·68
Wolverhampton	30,584	43,190	41·22	53,011	22·74	67,514	27·36
Bradford	13,264	16,012	20·73	26,307	64·30	43,527	65·46
Newcastle	33,048	32,573	− 1·44	41,794	28·31	53,613	28·29
Stoke-upon-Trent	23,278	31,557	35·57	40,237	27·51	51,589	28·21
Hull	29,580	37,005	25·10	44,520	20·31	51,911	16·60
Salford	18,088	24,744	36·80	32,600	31·75	50,810	55·86
Portsmouth	33,226	41,587	25·16	46,743	12·40	50,389	7·80
Oldham	21,766	29,479	35·43	38,201	29·59	50,513	32·23
Sunderland	24,998	25,821	3·29	31,891	23·51	40,735	27·73
Brighton	7,440	12,205	64·05	24,741	102·71	41,994	69·73
Total	1,583,958	1,898,173	19·84	2,362,282	24·45	3,009,841	27·41
Deduct London	958,863	1,138,815	18·77	1,378,947	21·09	1,654,994	20·02
Total (ex-London)	625,095	759,358	21·48	983,335	29·50	1,354,847	37·77

TOTAL OF LARGE

Total	3,615,245	4,247,796	17·50	5,212,075	22·70	6,394,167	22·68
Deduct London	958,863	1,138,815	18·77	1,378,947	21·09	1,654,994	20·02
Total (ex-London)	2,656,382	3,108,981	17·04	3,833,128	23·29	4,739,173	23·64

ENGLAND

Total	8,892,536	10,164,256	14·30	12,000,236	18·06	13,896,797	15·81
Deduct London	958,863	1,138,815	18·77	1,378,947	21·09	1,654,994	20·02
Total (ex-London)	7,933,673	9,025,441	13·76	10,621,289	17·68	12,241,803	15·26

    The predominating influence of the immense population of London as affecting the rates of increase of the aggregate town population in England and Wales is also very noticeable.

    In the case of some large towns, such as Wolverhampton, Newcastle-on-Tyne, Hull, Merthyr, Sunderland, and Preston, the period of maximum rate of increase occurred in the decade of 1831-41, while in the case of London itself, its period of maximum was reached in the next decade 1841-51. (*Vide* Appendix, Table G.)

Towns of England and Wales, between 1801 *and* 1871.

1841.		1851.		1861.		1871.		
Population.	Rate of Increase.	Population.	Rate of Increase.	Population.	Rate of Increase.	Population.	Rate of Increase.	
1,948,417	17·73	2,362,236	21·24	2,803,989	18·70	3,254,260	16·06	London
286,487	42·00	375,955	31·23	443,938	18·08	493,405	11·14	Liverpool
242,983	29·92	316,213	30·14	357,979	13·21	379,374	5·97	Manchester
182,922	27·04	232,841	27·29	296,076	27·16	343,787	16·11	Birmingham
152,074	23·24	172,270	13·28	207,165	20·26	259,212	25·12	Leeds
111,091	21·16	135,310	21·80	185,172	36·85	239,946	29·58	Sheffield
125,146	19·87	137,328	9·73	154,093	12·20	182,552	18·47	Bristol
93,245	38·11	119,748	28·42	147,670	23·32	156,978	6·30	Wolverhampton
66,715	53·27	103,778	55·55	106,218	2·35	145,830	37·29	Bradford
70,337	31·19	87,784	24·81	109,108	24·29	128,443	17·72	Newcastle
68,444	32·67	84,027	22·77	101,207	20·45	124,493	23·00	Stoke-upon-Trent
67,308	29·66	84,690	25·82	97,661	15·32	123,408	26·37	Hull
68,386	34·59	85,108	24·45	102,449	20·38	121,401	18·50	Salford
53,032	5·24	72,096	35·95	94,799	31·49	113,569	19·80	Portsmouth
60,451	19·67	72,357	19·70	94,344	30·39	113,100	19·88	Oldham
53,335	30·93	67,391	26·36	85,797	27·31	104,490	21·79	Sunderland
49,170	17·09	69,673	41·70	87,317	25·33	103,758	18·83	Brighton
3,699,543	22·92	4,578,805	23·77	5,474,982	19·57	6,388,006	16·68	Total
1,948,417	17·73	2,362,236	21·24	2,803,989	18·70	3,254,260	16·06	Deduct London
1,751,126	29·25	2,216,569	26·58	2,670,993	20·50	3,133,746	17·32	Total (ex-London)

AND SMALL TOWNS.

1841.		1851.		1861.		1871.		
7,679,737	20·10	9,213,942	19·98	10,717,260	16·32	12,576,626	17·32	Total
1,948,417	17·73	2,362,236	21·24	2,803,989	18·70	3,254,260	16·06	Deduct London
5,731,320	20·93	6,851,706	19·55	7,913,271	15·49	9,322,366	17·78	Total (ex-London)

AND WALES.

1841.		1851.		1861.		1871.		
15,909,132	14·48	17,927,609	12·69	20,066,224	11·93	22,712,266	13·19	Total
1,948,417	17·73	2,362,236	21·24	2,803,989	18·70	3,254,260	16·06	Deduct London
13,960,715	14·04	15,565,373	11·49	17,262,235	10·90	19,458,006	12·72	Total (ex-London)

Several remarkable instances are also to be met with of large
increments occurring in the rates of increase of town populations
during the last decade (1861-71) : notably in the case of Leicester,
Dudley, Derby, Rochdale, and some few other towns. It should
be observed, however, that the alteration of the boundary in some
other cases explains the higher rate of increase in the last decade.

Decrements in the Rates of Increase in Town Population.

The decrements in the rates of increase of the town population

c

are clearly indicated by the curved outlines on the diagrams which accompany this paper.* It is worthy of remark that in the case of the population of London, the decrements are very slight indeed, and the absence of the S shaped outlines of its population on the diagram, so conspicuous a feature in the population diagrams of most of the other large towns, shows that London has not yet reached that declining stage in the rate of its increase of population long since arrived at in the case of Liverpool, Manchester, and many other large towns.

The Population of Towns above Two Thousand and under Twenty Thousand Inhabitants.

The increase in the population of these towns has been much less rapid than in the case of the large towns, their aggregate population having been little more than doubled in the course of seventy years. Their maximum rate of increase, viz., 19 per cent., was reached, as in the case of the population of the rural districts, to which reference will presently be made, in the decade 1811-21; from that time down to the decade 1851-61, a rapid decrement of 84·10 per cent. occurred in this rate of increase, followed, however, in the last decade, 1861-71, by a somewhat remarkable rise of 51·44 per cent. The effect of this increase is clearly discernible in adding an increment of 6·13 per cent. to the rate of increase of the town population, and along with the increase observable in the case of the rural population in this decade, materially affecting the rate of increase of the entire population of England and Wales. (*Vide* Summary of Tables B to E, p. 7.)

Population of the Rural Districts.

The increase in the population of the rural districts of England and Wales during the first decade of this century was 12·11 per cent., or very similar to that of the smaller towns, and as in that case the maximum rate of increase (14·74) was reached in the following decade (1811-21), from that time down to the period of the census of 1851 the increase of the rural population was relatively very small, having in a period of thirty years only increased from 6,788,661 to 8,713,667, or 28·35 per cent., the decrement in the rate of increase during that time being rapid and continuous. From that period, however, up to 1871, there has been a rapid and continuous increment in the rate of increase, the effect of which, combined with that due to the increase of the population of the small towns, being such as to reduce in 1861 the decrement in the rate of increase of the entire population of England and Wales to 5·99 per cent., and in the following decade of 1871 to cause an increment of 10·56 per cent. in the rate of increase.

* See diagrams Plates 2 to 6.

The cause of the slow increase of the rural population between 1821 and 1851, is evidently in a great measure due to immigration into the towns; this will at once be seen on referring to the diagrams, and comparing the outlines showing the rates of increase of the town, rural, and total populations. It will not fail to be noticed that the periods of greatest increase in the town populations are coincident with those of greatest decrease in the case of the rural population. This is especially noticeable in the decade 1841-51, as may be seen from a reference to the first diagram on Plate 2, which shows that the aggregate population of the towns, which up to this period was considerably less than the population of the rural districts, equalled it about the middle of the decade, and at the end of the decade considerably exceeded it.

Future Increase of the Population of England and Wales.

It is unnecessary here to refer to the checks on the increase of population due to the limited area for food production in this country, as since Malthus's time, through the largely increased transit facilities afforded by the introduction of steam navigation, there is practically no limit to the area from which the food supplies of this country can be obtained, so long at least as those facilities for cheap and rapid communication with the great food producing countries of the world continue.

The rate of the future increase of the population of this country depends necessarily to a great extent upon the continuous growth of its trade and commerce, and upon the further development of that remarkable industrial activity which has been brought about during the last forty years, in a great measure by the agency of railways and steam navigation.

It cannot, however, be expected that there will be a repetition during the next forty years of the same rapid rate of commercial development which has been experienced during the last, and which has resulted from the creation as it were, in that short time, of an entirely new and rapid system of locomotion; still by means of the improvements which are continually being made in mechanical appliances, and the economies resulting therefrom, there is everything to indicate that the population of this country will continue to increase at the diminished or decremental rate which has occurred since the rate of increase of the population attained its maximum.

In the writer's opinion, in estimating the future population of England and Wales, it is putting it at its highest to assume a continuation of the 5·73 per cent. decrement, which during the last fifty years has obtained in the rate of increase of the population.

Estimate of the Census of 1881.

The following estimates of the population of England and Wales for the census year 1881 have been prepared by the writer on the basis indicated in this paper :—

Population of the large towns of 20,000 inhabitants and upwards, assumed to increase in 1871 at the average decremental rate of large towns, viz., 7·10 per cent.	11,771,491
Population of the small towns with over 2,000 and under 20,000 inhabitants, assuming the rate of increase of these towns during the last decade, 1871, viz., 11·07 per cent., to continue	3,082,970
Population of the rural districts, including small towns under 2,000 inhabitants, assuming the average decrement in the increment of the rate of increase of the last two decades, viz., 34·237 per cent.,* to continue	11,082,339†
Total population of England and Wales in 1881	25,936,800
The population of England and Wales, assuming the average decremental rate of increase of the total population during the last fifty years to continue, viz., 5·73 per cent.	25,535,000

In the opinion of the writer, the estimated increase of the population of the small towns, viz., 11·07 per cent., is too high, and probably the actual population will amount to a mean of the two results, viz., 25,735,900.

Future Increase of the Population of Great Britain in connection with the Question of the Coal Supply.

In the estimate of the future population of Great Britain prepared by the writer for the Commissioners appointed to inquire into the question of the coal supply, and given at p. xv of their report to the Queen, the average decrement in the rate of increase of the population which has obtained from the period when it

	Percentage Rate of Increase.	Increment per Cent.	Decrease in Increment per Cent.
* Rate of increase, 1841-51	5·88		
" '51-61	7·29	23·98	— 34·237
" '61-71	8·44	15·77	— 34·237
" '71-81	†9·315	10·37	

† Rural population 1871, 10,137,987 × 1·09315 = 11,082,339.

attained its maximum, 1811-21, down to the time of the last census of 1871, was employed as a means of determining the rates of increase in future decades.

In order more accurately to ascertain the rates of increase of the population during each decade of the present century, the Royal Commissioners included the army, navy, and the seamen belonging to the merchant service, which were omitted in the census returns prior to 1841, the military at home in that year being for the first time included, while persons on board vessels in the navy or merchant service, and lying in harbours, creeks, and rivers, were for the first time included with the military in the later census returns of 1851, 1861 and 1871.*

The population, however, of Great Britain for 1871, given in the before-mentioned tables in the Coal Commissioners' report, and which was furnished to them just prior to the publication of the preliminary census report of 1871, proves to have been somewhat short of the actual numbers † given in the complete census returns of 1871; in addition to this, the army, navy, and seamen employed in the merchant service have not, as in the case of the previous decades, been included in these tables for the decade of 1871; the effect of this both upon the rate of increase of the population during that particular decade, and also upon the average decrement in the rate of the increase of the population during the period of fifty years, being very appreciable (see Tables H and I).

An amended estimate of the future increase of the population of Great Britain, based on these corrected figures, is given in Table L in the Appendix, p. 33, and graphically shown on diagram, Plate 8.

Estimates of the future increase of the populations of England and Wales, and of Scotland, are given separately in Tables M and N in the Appendix, pp. 34 and 35, the average decrement in the rates of increase of the population of each kingdom during the last fifty years being employed in each case in determining the rates of increase in the future decades.

The combined results obtained by thus separately estimating the future increase of the respective populations of the two kingdoms during a long period of years, are somewhat in excess of that obtained in the case of the aggregate population of the two kingdoms; in the latter case the initial rate of increase is considerably less, and the rate of decrement more rapid.

The effect of this combination of the populations of the two kingdoms at once reduces the initial rate of increase of the total

---

* See foot note p. x, Summary Table Census 1871, vol. i.

† Population, Great Britain 26,072,284
    ,,   Coal Commissioners' report 26,062,721

                Difference 9,563

population in 1871 from 12·990 per cent., in the case of England
and Wales, to 12·533 per cent. when combined with that of
Scotland, and at the same time it increases the decrement in the
rate of increase from 4·595 per cent. to 4·740 per cent. per
decade.

The conclusions which, in the opinion of the writer, are to be
drawn from these facts are, that the larger the area of population
dealt with, the more accurate and reliable will be any estimates
which are made to determine the future increase of that population.

The decrement in the rate of increase of the population of
London which has only obtained during the two last decades
scarcely affords, in the opinion of the writer, sufficient data for
estimating its future increase for any lengthened period, and that
given in Table O in the Appendix, and diagram Plate 9, has been
made rather with the object of showing the unreliableness of any
such enormous estimates as those which have recently appeared in
connection with the question of the water supply of the metropolis,
where the population in the course of the next century is estimated
at over 17 millions.

Table showing the Population of England and Wales, Scotland and Great Britain
Islands in the
[Extracted from the Census Returns for

Decade.	England and Wales.			Scotland.		
	Population.	Rate of Increase per Decade.	Decrement in Rate of Increase.	Population.	Rate of Increase per Decade.	Decrement in Rate of Increase.
	No.	Per cnt.	Per cnt.	No.	Per cnt.	Per cnt.
1801........	9,156,171	14·180		1,678,452	12·248	
11........	10,454,529	* 16·434		1,884,044	* 13·444	— 6·627
'21......	12,172,664	15·438	— 6·060	2,137,325	12·553	— 18·291
'31......	14,051,986	14·113	— 8·582	2,405,610	10·257	— 0·741
'41......	16,035,198	12·591	— 2·725	2,652,339	10·181	
'51......	18,054,170	12·043	— 2·725	2,922,362	5·969	— 3·148
'61......	20,228,497	* 12·990	— 2·725	3,096,808	* 9·550	— 3·148
'71	22,856,164			3,392,559		
		5) — 22·817			5) — 31·955	
Average diminution of rate per decade		— 4·563		Average diminution of rate per decade	— 6·391	
*Initial and terminal periods............		— 4·595		* Initial and terminal periods	— 6·610	

†This population was subsequently given in the completed census, published in 1873, as

Future Increase of the Population of London.

The population of London, and its remarkable growth during the present century, affords in itself the subject of a paper, and as a small contribution towards this the writer has prepared certain tables and diagrams, viz.:—

1st. Sectional Diagrams showing the population and rates of increase or decrease of population in each district, sub-district, or parish within the registrar-general's district from 1801 to 1871.‡

2nd. Diagram map, showing the rates of increase or decrease of population in each district, sub-district, or parish within the registrar-general's district from 1801 to 1871.‡

3rd. Diagram map, showing the number of persons per acre in each district, sub-district, or parish within the registrar-general's district, from 1801 to 1871.‡

4th. Diagram map, showing the number of persons per acre in population in each district, sub-district, or parish within the registrar-general's district from 1801 to 1871.‡

5th. Estimate of the prospective increase of the population of London. (Table O in the Appendix.)

(including the Army, Navy, Marines, and Merchant Seamen, but excluding the British Seas.)
1871, general table, vol. iv, Table 4, p. 5.]

Great Britain.			Great Britain. Coal Commissioners' Report.	(Table No. II, p. xv, Report 1871).		Decade.
Population.	Rate of Increase per Decade.	Decrement in Rate of Increase.	Population.	Rate of Increase per Decade.	Decrement in Rate of Increase.	
No.	Per cnt.	Per cnt.	No.	Per cnt.	Per cnt.	
10,834,623						1801
	13·881					
12,338,573			12,338,573			'11
	*15·977			15·977		
14,309,989		− 6·065	14,309,989		− 6·065	'21
	15·008			15·008		
16,457,596		− 9·721	16,457,596		− 9·721	'31
	13·549			13·549		
18,687,537			18,687,537		− 9·595	'41
	12·249	− 2·562		12·249		
20,976,532			20,976,532		− 8·588	'51
	11·197	− 2·562		11·197		
23,325,305			23,325,305		+ 4·814	'61
	*12·533	− 2·562		11·736		
26,248,723			26,062,721†			'71
		5) − 23·472			− 33·969 + 4·814	
Average diminution of rate per decade		− 4·694			5) − 29·155	
*Initial and terminal periods		− 4·740	Average diminution of rate per decade............		− 5·831	

(In the Decrement in Rate of Increase column: Average rate between 1841 and 1871.)

26,072,284; the army and navy, &c., 176,439, being omitted;—total, 26,248,723.

‡ Not given with this paper.

APPENDIX.*   TABLE B.—*England and Wales. Aggregate Population, and Increase or*

County.	1801.	1811.	Increase.	1821.	Increase.	1831.	Increase.
	No.	No.	Per cnt.	No.	Per cnt.	No.	Per cnt
Bedford	3,095	3,716	20·07	4,529	21·88	5,693	25·70
Berks	9,742	10,788	10·74	12,867	19·27	15,595	21·20
Buckingham	16,993	18,435	8·49	21,717	17·80	24,162	11·26
Cambridge	10,087	11,108	10·12	14,142	27·31	20,917	47·91
Chester	39,414	46,779	18·69	60,734	29·83	74,137	22·07
Cornwall	Nil	Nil	—	Nil	—	Nil	—
Cumberland	9,415	11,476	21·89	14,416	25·62	18,865	30·86
Derby	10,832	13,043	20·41	17,423	33·58	23,627	35·61
Devon	61,444	76,306	24·19	86,616	13·51	107,358	23·95
Dorset	Nil	Nil	—	Nil	—	Nil	—
Durham	53,212	59,014	10·90	70,867	20·09	90,927	28·31
Essex	15,430	17,449	13·08	19,808	14·04	23,158	16·38
Gloucester	91,961	108,206	17·67	134,844	24·61	167,282	24·06
Hereford	Nil	Nil	—	Nil	—	Nil	—
Hertford	„	„	—	„	—	„	—
Huntingdon	„	„	—	„	—	„	—
Kent	96,024	123,257	28·36	141,722	14·98	162,296	14·52
Lancaster	304,855	385,271	26·37	515,780	33·08	721,805	39·95
Leicester	17,005	23,453	37·92	31,036	32·33	40,639	30·94
Lincoln	10,340	13,141	27·09	15,178	15·54	17,806	17·31
Middlesex	751,753	874,995	16·39	1,052,319	20·27	1,252,967	19·07
Monmouth	1,423	3,025	112·58	4,951	63·67	7,062	42·64
Norfolk	53,427	56,704	6·13	71,295	25·73	85,651	20·14
Northampton	7,020	8,427	20·04	10,793	28·07	15,351	42·24
Northumberland	51,078	55,409	8·48	70,820	27·81	83,396	17·76
Nottingham	28,801	34,030	18·15	40,190	18·10	50,220	24·96
Oxford	11,694	12,931	10·58	16,364	26·55	20,649	26·19
Rutland	Nil	Nil	—	Nil	—	Nil	—
Salop	31,043	33,630	8·34	37,119	10·37	38,732	4·34
Somerset	33,196	38,408	15·70	46,700	21·59	50,800	8·78
Southampton	41,139	51,204	24·47	60,096	17·37	69,713	16·00
Stafford	72,759	96,653	32·84	118,218	22·31	150,888	27·64
Suffolk	11,277	13,670	21·22	17,186	25·72	20,201	17·54
Surrey	161,642	202,941	25·55	259,714	27·97	327,820	26·33
Sussex	26,526	34,743	30·98	53,574	54·20	77,447	44·56
Warwick	87,019	101,219	16·32	125,353	23·84	177,493	41·60
Westmoreland	Nil	Nil	—	Nil	—	Nil	—
Wilts	22,139	23,777	7·40	27,028	13·67	29,118	7·73
Worcester	28,370	36,492	28·63	46,678	27·91	58,076	24·42
York, E., N., and W.R.	211,884	248,974	17·51	322,486	29·53	437,064	35·53
Total	2,382,039	2,848,674	19·59	3,542,653	24·36	4,466,915	26·09
WALES.							
Anglesey	Nil	Nil	—	Nil	—	Nil	—
Brecon	„	„	—	„	—	„	—
Cardigan	„	„	—	„	—	„	—
Carmarthen	„	„	—	„	—	„	—
Carnarvon	„	„	—	„	—	„	—
Denbigh	„	„	—	„	—	„	—
Flint	„	„	—	„	—	„	—
Glamorgan	22,114	29,365	32·79	39,376	34·09	53,140	34·96
Merioneth	Nil	Nil	—	Nil	—	Nil	—
Montgomery	„	„	—	„	—	„	—
Pembroke	„	„	—	„	—	„	—
Radnor	„	„	—	„	—	„	—
Total	22,114	29,365	32·79	39,376	34·09	53,140	34·96
Total England & Wales	2,404,153	2,878,039	19·71	3,582,029	24·46	4,520,055	26·19

* Tables A¹ to A³³, containing particulars of the population, rates of increase, &c., of the town and

Decrease per Cent. of the Large Towns of over 20,000 Inhabitants between 1801 and 1871.

1841. No.	Increase Per cnt.	1851. No.	Increase Per cnt.	1861. No.	Increase Per cnt.	1871. No.	Increase Per cnt.	County.
7,748	36·10	12,787	65·04	17,821	39·37	20,733	16·34	Bedford
18,937	21·43	21,456	13·30	25,045	16·73	32,313	29·02	Berks
25,337	4·86	26,794	5·75	27,090	1·10	28,760	6·16	Buckingham
24,453	16·90	27,815	13·75	26,361	−5·23	33,996	28·96	Cambridge
118,212	59·45	175,878	48·78	198,642	12·94	210,784	6·11	Chester
Nil	—	Nil	—	Nil	—	Nil	—	Cornwall
21,550	14·23	26,310	22·09	29,417	11·81	31,049	5·55	Cumberland
32,741	38·57	40,609	24·03	43,091	6·11	61,381	42·45	Derby
123,265	14·82	154,542	25·37	185,550	21·17	200,008	7·79	Devon
Nil	—	Nil	—	Nil	—	Nil	—	Dorset
117,290	28·99	143,225	22·12	183,769	28·30	253,271	37·82	Durham
25,480	10·13	30,029	17·85	39,803	32·55	49,629	24·69	Essex
194,549	16·30	208,914	7·38	229,303	9·76	265,681	45·86	Gloucester
Nil	—	Nil	—	Nil	—	Nil	—	Hereford
,,	—	,,	—	,,	—	,,	—	Hertford
,,	—	,,	—	,,	—	,,	—	Huntingdon
208,228	28·30	253,864	21·91	339,460	33·72	392,377	15·59	Kent
966,797	33·94	1,239,702	28·23	1,475,422	19·01	1,713,250	16·12	Lancaster
50,806	25·02	60,584	19·25	68,056	12·33	95,220	39·91	Leicester
20,594	15·66	29,796	44·68	36,059	21·02	53,748	49·06	Lincoln
1,453,406	15·99	1,755,429	20·78	2,044,772	16·49	2,306,800	12·81	Middlesex
10,815	53·15	19,323	78·68	23,249	20·32	27,069	16·43	Monmouth
90,209	5·32	99,592	10·40	109,701	10·15	122,205	11·40	Norfolk
21,242	38·38	26,657	25·49	32,813	23·10	45,080	37·39	Northampton
102,913	23·40	126,966	23·08	156,923	23·88	197,623	25·94	Northumberland
52,164	3·87	57,407	10·05	74,693	30·11	86,621	15·97	Nottingham
24,258	17·48	27,843	14·78	27,560	−1·02	31,404	13·95	Oxford
Nil	—	Nil	—	Nil	—	Nil	—	Rutland
39,534	2·07	43,692	10·52	47,374	8·43	44,614	−5·83	Salop
53,196	4·72	54,240	1·96	52,528	−3·16	53,714	2·26	Somerset
80,776	15·87	107,401	32·96	141,759	31·99	167,310	18·02	Southampton
203,389	34·79	269,105	32·31	340,558	26·55	392,575	15·27	Stafford
25,384	25·66	32,914	29·66	37,950	15·30	42,947	13·17	Suffolk
399,247	21·79	482,435	20·84	579,748	20·17	742,155	28·01	Surrey
89,425	15·46	117,237	31·10	142,849	21·85	175,079	22·56	Sussex
226,818	27·79	285,377	25·82	355,125	24·44	406,045	14·34	Warwick
Nil	—	Nil	—	Nil	—	Nil	—	Westmoreland
35,409	21·61	35,503	0·27	36,893	3·91	43,622	18·24	Wilts
73,663	26·84	83,952	13·97	91,601	9·11	141,179	54·12	Worcester
556,628	27·36	694,731	24·81	838,789	20·74	1,144,359	36·43	York, E., N., and W. R.
5,494,463	23·00	6,772,109	23·25	8,059,774	19·01	9,612,601	19·27	**Total**
								WALES.
Nil	—	Nil	—	Nil	—	Nil	—	Anglesey
,,	—	,,	—	,,	—	,,	—	Brecon
,,	—	,,	—	,,	—	,,	—	Cardigan
,,	—	,,	—	,,	—	,,	—	Carmarthen
,,	—	,,	—	,,	—	,,	—	Carnarvon
,,	—	,,	—	,,	—	,,	—	Denbigh
,,	—	,,	—	,,	—	,,	—	Flint
77,712	46·24	112,892	45·27	158,435	40·35	188,286	18·84	Glamorgan
Nil	—	Nil	—	Nil	—	Nil	—	Merioneth
,,	—	,,	—	,,	—	,,	—	Montgomery
,,	—	,,	—	,,	—	,,	—	Pembroke
,,	—	,,	—	,,	—	,,	—	Radnor
77,712	46·24	112,892	45·27	158,435	40·35	188,286	18·84	**Total**
5,572,175	23·28	6,885,001	23·56	8,218,209	19·36	9,800,887	19·25	**Total England & Wales**

rural, and total population of each county in England and Wales from 1801 to 1871, are not printed.

TABLE C.—*England and Wales. Aggregate Population, and Increase or Decrease per*

County.	1801.	1811.	Increase.	1821.	Increase.	1831.	Increase.
	No.	No.	Per cnt.	No.	Per cnt.	No.	Per cnt.
Bedford	10,235	11,529	12·64	14,351	24·48	17,320	20·69
Berks	28,895	31,520	9·09	35,092	11·33	39,299	11·99
Buckingham	22,021	25,472	15·67	28,260	10·95	31,214	10·45
Cambridge	13,786	15,540	12·72	19,079	22·72	22,714	19·05
Chester	17,775	21,688	22·01	28,419	31·04	39,668	39·58
Cornwall	46,407	53,025	14·26	65,634	23·78	80,049	21·96
Cumberland	35,010	40,148	14·68	48,662	21·21	51,333	5·49
Derby	21,220	28,053	20·81	35,038	24·90	41,692	18·99
Devon	59,993	65,702	9·52	78,352	19·25	89,068	13·68
Dorset	35,539	39,470	11·06	47,333	19·92	53,548	15·13
Durham	14,694	15,400	4·80	19,639	27·53	21,860	11·31
Essex	28,339	31,561	11·37	36,456	15·51	41,898	14·93
Gloucester	23,974	26,672	11·26	30,340	13·75	35,202	16·02
Hereford	15,618	16,433	5·22	19,710	19·94	23,277	18·10
Hertford	21,676	25,165	16·10	30,394	20·78	34,476	13·43
Huntingdon	10,926	12,759	16·78	14,575	14·23	16,514	13·30
Kent	50,345	64,566	28·25	70,439	9·10	81,985	16·39
Lancaster	70,352	83,988	19·40	105,365	25·44	125,849	19·44
Leicester	19,383	22,601	16·60	26,487	17·20	32,763	23·69
Lincoln	42,242	50,053	18·49	61,380	22·63	70,522	14·89
Middlesex	3,861	4,453	15·33	4,707	5·70	5,529	17·46
Monmouth	8,374	9,408	12·35	11,148	18·50	13,127	17·75
Norfolk	30,282	31,813	5·06	37,997	19·44	43,832	15·36
Northampton	18,586	20,193	8·65	23,596	16·85	26,172	10·92
Northumberland	14,673	16,306	11·13	18,389	12·77	19,685	7·05
Nottingham	18,295	20,439	11·72	23,576	15·35	27,943	18·52
Oxford	20,417	22,188	8·68	25,373	14·35	27,421	8·07
Rutland	3,055	3,203	4·85	3,790	18·33	4,147	9·42
Salop	45,291	48,512	7·11	52,333	7·88	58,991	12·72
Somerset	35,637	38,134	7·01	45,103	18·27	51,678	14·58
Southampton	39,785	43,149	8·46	51,045	18·30	57,643	12·93
Stafford	42,539	45,505	12·25	53,644	17·89	59,148	10·26
Suffolk	21,886	23,449	7·14	27,732	18·27	31,213	12·55
Surrey	26,497	29,599	11·71	34,320	15·95	39,825	16·04
Sussex	29,040	37,559	29·34	44,365	18·12	50,157	13·06
Warwick	21,093	23,144	9·72	28,789	24·39	33,441	16·16
Westmoreland	8,015	8,793	9·71	10,438	18·71	11,577	10·91
Wilts	64,426	66,704	3·54	79,740	19·54	86,629	8·64
Worcester	12,697	13,494	6·28	15,099	11·89	17,076	13·09
York, E., N., & W. R.	86,722	98,487	13·57	117,072	18·90	129,947	11·00
Total	1,139,601	1,285,877	12·84	1,523,261	18·46	1,745,432	14·58
WALES.							
Anglesey	6,283	7,047	12·16	9,364	32·88	10,552	12·69
Brecon	1,583	2,247	41·95	2,906	29·33	3,343	15·04
Cardigan	5,989	6,695	11·79	8,758	30·81	10,162	16·03
Carmarthen	8,112	10,487	29·28	12,997	23·93	16,135	24·15
Carnarvon	8,620	10,618	23·18	13,666	28·71	16,919	23·80
Denbigh	9,755	10,672	9·40	12,071	13·11	13,695	13·45
Flint	6,384	7,393	15·81	9,300	25·79	11,008	18·37
Glamorgan	5,477	6,059	10·63	6,698	10·55	9,082	35·59
Merioneth	Nil	Nil	—	Nil	—	Nil	—
Montgomery	9,355	11,756	25·67	15,596	32·66	19,543	25·31
Pembroke	8,094	9,069	12·05	13,331	46·99	15,883	19·14
Radnor	1,839	1,837	−0·10	2,098	14·21	2,358	12·39
Total	71,491	83,880	17·33	106,785	27·31	128,680	20·50
Total England & Wales	1,211,092	1,369,757	13·10	1,630,046	19·00	1,874,112	14·98

Cent. of the Small Towns of 2,000 *and under* 20,000 *Inhabitants between* 1801 *and* 1871.

1841.	Increase	1851.	Increase	1861.	Increase.	1871.	Increase	County.
No.	Per cnt.	No.	Per cnt.	No.	Per cnt	No.	Per cnt.	
21,532	24·32	26,168	21·53	29,540	12·89	34,339	16·25	Bedford
43,985	11·93	45,065	2·46	46,659	3·54	52,959	13·50	Berks
32,979	5·66	34,054	3·26	33,620	−1·27	35,286	4·95	Buckingham
25,747	13·35	29,960	16·36	26,882	−10·27	27,356	1·76	Cambridge
48,595	22·50	54,932	13·04	64,098	16·69	67,963	6·03	Chester
98,180	22·65	108,510	10·52	115,354	6·31	118,537	2·76	Cornwall
53,720	4·61	60,232	12·12	62,708	4·11	65,175	3·93	Cumberland
53,079	27·31	63,048	18·78	77,448	22·84	79,090	2·12	Derby
96,235	8·05	102,290	6·29	101,762	−0·52	106,930	5·08	Devon
58,362	8·99	63,003	7·95	64,406	2·23	66,920	3·90	Dorset
29,073	33·00	39,066	34·37	45,843	17·35	51,443	12·21	Durham
46,928	12·01	54,105	15·29	57,585	6·34	64,449	12·02	Essex
37,442	6·36	39,847	6·42	38,148	−4·26	40,925	7·28	Gloucester
24,577	5·58	25,963	5·64	31,187	20·12	33,272	6·69	Hereford
40,794	18·33	46,259	13·40	49,282	6·53	53,706	8·98	Hertford
17,969	8·81	19,930	10·91	19,908	−0·11	20,151	1·22	Huntingdon
90,816	10·26	104,815	15·41	117,966	12·55	146,302	24·02	Kent
149,848	19·07	171,360	14·35	206,954	20·77	248,163	19·91	Lancaster
33,760	3·04	35,207	4·29	34,993	−0·61	37,605	7·46	Leicester
82,630	17·17	93,036	12·59	92,038	−1·07	98,192	6·69	Lincoln
5,706	3·20	5,813	1·88	5,985	2·96	7,023	17·34	Middlesex
14,141	7·72	15,548	9·95	15,324	−1·44	15,544	1·44	Monmouth
48,849	11·44	55,280	13·16	51,539	−6·17	52,875	2·59	Norfolk
29,546	12·89	31,647	7·11	36,491	15·31	46,093	26·31	Northampton
19,204	−2·44	22,413	16·71	20,615	−8·02	20,328	−1·39	Northumberland
29,690	6·25	32,688	10·10	33,570	2·70	37,633	12·10	Nottingham
32,105	17·08	32,484	1·18	33,801	4·05	35,803	5·92	Oxford
4,760	14·78	5,099	7·12	5,145	0·90	5,690	10·59	Rutland
64,086	8·64	65,027	1·47	69,919	7·52	73,780	5·52	Salop
58,207	12·63	60,294	3·59	60,941	1·07	63,310	3·89	Somerset
68,083	18·11	78,764	15·69	85,620	8·71	101,435	18·47	Southampton
68,265	15·41	77,547	13·60	87,080	12·29	94,747	8·81	Stafford
33,517	7·38	36,196	7·99	35,251	−2·61	37,804	7·24	Suffolk
46,332	16·34	52,833	14·03	68,786	30·19	91,356	32·81	Surrey
54,979	9·62	57,730	5·00	59,606	3·25	71,221	19·49	Sussex
35,044	4·79	41,754	19·15	42,939	2·84	43,363	0·99	Warwick
11,519	−0·50	11,829	2·69	12,029	1·69	13,446	11·78	Westmoreland
91,567	5·70	89,923	−1·80	87,947	−2·20	90,987	3·46	Wilts
18,291	7·11	19,019	3·98	18,850	−0·89	22,012	16·78	Worcester
142,049	9·31	156,073	9·87	174,436	11·77	196,980	12·92	York, E., N., & W. R.
1,962,191	12·42	2,164,811	10·32	2,322,205	7·27	2,570,193	10·68	Total
								WALES.
10,396	−1·48	12,752	22·66	13,275	4·10	13,672	2·99	Anglesey
5,317	59·05	6,070	14·16	5,639	−7·10	6,291	11·56	Brecon
11,296	11·16	11,760	4·11	11,646	−0·97	14,485	24·38	Cardigan
18,053	11·89	21,161	17·22	23,294	10·08	27,630	18·62	Carmarthen
19,003	12·32	22,210	16·88	22,907	3·14	27,540	20·23	Carnarvon
15,450	12·82	16,614	7·54	17,888	7·67	20,223	13·05	Denbigh
14,588	32·52	14,509	−0·54	14,561	0·36	18,958	30·20	Flint
11,140	22·66	15,302	37·36	18,027	17·81	26,359	46·22	Glamorgan
Nil	—	Nil	—	Nil	—	Nil	—	Merioneth
19,700	0·80	18,901	−4·06	19,268	1·94	19,480	1·10	Montgomery
17,950	13·01	22,508	25·39	28,079	24·75	28,718	2·28	Pembroke
2,478	5·09	2,345	−5·37	2,262	−3·54	2,190	−3·18	Radnor
145,371	12·97	164,130	12·90	176,846	7·75	205,546	16·23	Total
2,107,562	12·46	2,328,941	10·50	2,499,051	7·31	2,775,739	11·07	Total England & Wales

TABLE D.—*England and Wales. Aggregate Population, and Increase or Decrease*

County.	1801.	1811.	Increase.	1821.	Increase.	1831.	Increase.
	No.	No.	Per cnt.	No.	Per cnt.	No.	Per cnt.
Bedford	13,330	15,245	14·37	18,880	23·84	23,013	21·89
Berks	38,637	42,308	9·50	47,959	13·36	54,894	14·46
Buckingham	39,014	43,907	12·55	49,977	13·83	55,376	10·80
Cambridge	23,873	26,648	11·62	33,221	24·67	43,631	31·33
Chester	57,189	68,467	19·72	89,153	30·21	113,805	27·65
Cornwall	46,407	53,025	14·26	65,634	23·78	80,049	21·96
Cumberland	44,425	51,624	16·20	63,078	22·19	70,218	11·32
Derby	34,052	41,096	20·67	52,461	27·67	65,319	24·51
Devon	121,437	142,008	16·94	164,968	15·82	196,426	19·44
Dorset	35,539	39,470	11·06	47,333	19·92	53,548	13·13
Durham	67,906	74,414	9·58	90,506	21·62	112,787	24·62
Essex	43,769	49,010	11·97	56,354	14·90	65,056	15·44
Gloucester	115,935	134,878	16·33	165,184	22·46	202,484	22·58
Hereford	15,618	16,433	5·22	19,710	19·94	23,277	18·10
Hertford	21,676	25,165	16·10	30,394	20·78	34,476	13·43
Huntingdon	10,926	12,759	16·78	14,575	14·23	16,514	13·30
Kent	146,369	187,823	28·32	212,161	12·96	244,281	15·14
Lancaster	375,207	469,259	25·07	621,145	32·37	847,654	36·46
Leicester	36,388	46,054	26·56	57,523	24·90	73,402	27·61
Lincoln	52,582	63,194	20·18	76,558	21·15	88,328	15·37
Middlesex	755,614	879,448	16·39	1,057,026	20·19	1,258,496	19·06
Monmouth	9,797	12,433	26·91	16,099	29·49	20,189	25·41
Norfolk	83,709	88,517	5·74	109,292	23·47	129,483	18·47
Northampton	25,606	28,620	11·77	34,389	20·16	41,523	20·75
Northumberland	65,751	71,715	9·07	89,209	24·39	103,081	15·55
Nottingham	47,096	54,469	15·66	63,766	17·07	78,163	22·58
Oxford	32,111	35,119	9·37	41,737	18·84	48,070	15·17
Rutland	3,055	3,203	4·85	3,790	18·33	4,147	9·42
Salop	76,334	82,142	7·61	89,452	8·90	97,723	9·25
Somerset	68,833	76,542	11·20	91,803	19·94	102,478	11·63
Southampton	80,924	94,353	16·60	111,141	17·79	127,356	14·59
Stafford	113,298	142,158	25·48	171,862	20·89	210,036	22·22
Suffolk	33,163	37,119	11·93	44,918	21·01	51,414	14·46
Surrey	188,139	232,540	23·60	294,034	26·44	367,645	25·04
Sussex	55,566	72,302	30·12	97,939	35·46	127,604	30·29
Warwick	108,112	124,363	15·03	154,142	23·95	210,934	36·84
Westmoreland	8,015	8,793	9·71	10,438	18·71	11,577	10·91
Wilts	86,565	90,481	4·53	106,768	18·00	115,747	8·41
Worcester	41,067	49,986	21·72	61,777	23·59	75,152	21·65
York, E., N., and W. R.	298,606	347,461	16·36	439,558	26·51	567,011	28·99
Total	3,521,640	4,134,551	17·41	5,065,914	22·52	6,212,347	22·63
WALES.							
Anglesey	6,283	7,047	12·16	9,364	32·88	10,552	12·69
Brecon	1,583	2,247	41·95	2,906	29·33	3,343	15·04
Cardigan	5,989	6,695	11·79	8,758	30·81	10,162	16·03
Carmarthen	8,112	10,487	29·28	12,997	23·93	16,135	24·15
Carnarvon	8,620	10,618	23·18	13,666	28·71	16,919	23·80
Denbigh	9,755	10,672	9·40	12,071	13·11	13,695	13·45
Flint	6,384	7,393	15·81	9,300	25·79	11,008	18·37
Glamorgan	27,591	35,424	28·39	46,074	30·06	62,222	35·05
Merioneth	Nil	Nil	—	Nil	—	Nil	—
Montgomery	9,355	11,756	25·67	15,596	32·66	19,543	25·31
Pembroke	8,094	9,069	12·05	13,331	46·99	15,883	19·14
Radnor	1,839	1,837	−0·10	2,098	14·21	2,358	12·39
Total	93,605	113,245	20·98	146,161	29·07	181,820	24·40
Total England & Wales	3,615,245	4,247,796	17·50	5,212,075	22·70	6,394,167	22·68

per Cent. of the Large and Small Towns combined, between 1801 and 1871.

1841. No.	Increase Per cnt.	1851. No.	Increase Per cnt	1861. No.	Increase Per cnt.	1871. No.	Increase Per cnt.	County.
29,280	27·23	38,955	33·04	47,361	21·58	55,072	16·28	Bedford
62,922	14·62	66,521	5·72	71,704	7·79	85,272	18·92	Berks
58,316	5·31	60,848	4·34	60,710	-0·23	64,046	5·49	Buckingham
50,200	15·06	57,775	15·09	53,243	-7·84	61,352	15·23	Cambridge
166,807	46·58	230,810	38·37	262,740	13·83	278,747	6·09	Chester
98,180	22·65	108,510	10·52	115,354	6·31	118,537	2·76	Cornwall
75,270	7·19	86,542	14·98	92,125	6·45	96,224	4·45	Cumberland
85,820	31·39	103,657	20·78	120,539	16·29	140,471	16·53	Derby
219,500	11·74	256,832	16·10	287,312	12·75	306,938	6·83	Devon
58,362	8·99	63,003	7·95	64,406	2·23	66,920	3·90	Dorset
146,363	29·77	182,291	24·55	229,612	25·96	304,714	32·71	Durham
72,408	11·30	84,134	16·19	97,338	13·07	114,078	44·92	Essex
231,991	14·58	248,761	7·23	267,451	7·51	306,606	14·64	Gloucester
24,577	5·58	25,963	5·64	31,187	20·12	33,272	6·69	Hereford
40,794	18·33	46,259	13·40	49,282	6·53	53,706	8·98	Hertford
17,969	8·81	19,930	10·91	19,908	-0·11	20,151	1·22	Huntingdon
299,044	22·42	358,679	19·94	457,426	27·53	538,679	17·76	Kent
1,116,645	31·73	1,411,062	26·37	1,682,376	19·23	1,961,413	16·58	Lancaster
84,566	15·21	95,791	13·27	103,049	7·58	132,825	28·90	Leicester
103,224	16·86	122,832	18·99	128,097	4·29	151,940	18·61	Lincoln
1,459,112	15·94	1,761,242	20·70	2,050,757	16·44	2,313,823	12·82	Middlesex
24,956	23·61	34,871	39·73	38,573	10·62	42,613	10·47	Monmouth
139,058	7·40	154,872	11·37	161,240	4·11	175,080	8·58	Norfolk
50,788	22·31	58,304	14·80	69,304	18·87	91,173	31·56	Northampton
122,117	18·47	149,379	22·32	177,538	18·85	217,951	22·76	Northumberland
81,854	4·72	90,095	10·07	108,263	20·17	124,254	14·77	Nottingham
56,363	17·25	60,327	7·03	61,361	1·71	67,207	9·53	Oxford
4,760	14·78	5,099	7·12	5,145	0·90	5,690	10·59	Rutland
103,620	6·03	108,719	4·92	117,293	7·89	118,394	0·94	Salop
111,403	8·71	114,534	2·81	113,469	-0·93	117,024	3·13	Somerset
148,859	16·88	186,165	25·06	227,379	22·14	268,745	18·20	Southampton
271,654	29·33	346,652	27·61	427,638	23·36	487,322	13·96	Stafford
58,901	14·56	69,110	17·33	73,201	5·92	80,751	10·31	Suffolk
445,579	21·20	535,268	20·13	648,534	21·16	833,511	28·52	Surrey
144,404	13·17	174,967	21·17	202,455	15·71	246,300	21·66	Sussex
261,862	24·15	327,131	24·93	398,064	21·68	449,408	12·90	Warwick
11,519	-0·50	11,829	2·69	12,029	1·69	13,446	11·78	Westmoreland
126,976	9·71	125,426	-1·22	124,840	-0·47	134,609	7·83	Wilts
91,954	22·36	102,971	11·98	110,451	7·26	163,191	47·75	Worcester
698,677	23·22	850,804	21·77	1,013,225	19·09	1,341,339	32·38	York, E., N., & W. R.
7,456,654	20·03	8,936,920	19·85	10,381,979	16·17	12,182,794	17·35	Total
								WALES.
10,396	-1·48	12,752	22·66	13,275	4·10	13,672	2·99	Anglesey
5,317	59·05	6,070	14·16	5,639	-7·10	6,291	11·56	Brecon
11,296	11·16	11,760	4·11	11,646	-0·97	14,485	24·38	Cardigan
18,053	11·89	21,161	17·22	23,294	10·08	27,630	18·62	Carmarthen
19,003	12·32	22,210	16·88	22,907	3·14	27,540	20·23	Carnarvon
15,450	12·82	16,614	7·54	17,888	7·67	20,223	13·05	Denbigh
14,588	32·52	14,509	-0·54	14,561	0·36	18,958	30·20	Flint
88,852	42·80	128,194	44·27	176,462	37·65	214,645	21·64	Glamorgan
Nil	—	Nil	—	Nil	—	Nil	—	Merioneth
19,700	0·80	18,901	-4·06	19,268	1·94	19,480	1·10	Montgomery
17,950	13·01	22,508	25·39	28,079	24·75	28,718	2·28	Pembroke
2,478	5·09	2,345	-5·37	2,262	-3·54	2,190	-3·18	Radnor
223,083	22·69	277,022	24·18	335,281	21·03	393,832	17·46	Total
7,679,737	20·10	9,213,942	19·98	10,717,260	16·32	12,576,626	17·35	Total England & Wales

TABLE E.—*England and Wales.   Aggregate Population and Increase*

County.	1801.	1811.	Increase.	1821.	Increase.	1831.	Increase.
	No.	No.	Per cnt.	No.	Per cnt.	No.	Per cnt.
Bedford	50,063	54,968	9·80	65,172	14·57	72,470	11·20
Berks	71,843	77,122	7·35	84,680	9·80	91,340	7·86
Buckingham	69,118	74,158	7·29	85,156	14·83	91,601	7·57
Cambridge	65,473	74,461	13·73	89,166	19·75	100,324	12·51
Chester	135,116	158,564	17·35	180,945	14·12	220,586	21·91
Cornwall	145,874	167,500	14·83	195,411	16·66	221,257	13·23
Cumberland	72,805	82,041	12·69	93,046	13·41	99,044	6·45
Derby	127,515	144,391	13·24	161,190	11·63	171,851	6·61
Devon	218,871	240,770	10·01	273,949	13·78	297,482	8·59
Dorset	78,913	85,248	8·03	97,597	14·49	105,837	8·44
Durham	81,478	90,879	11·54	103,005	13·34	126,469	22·78
Essex	183,913	203,463	10·63	233,070	14·55	252,451	8·32
Gloucester	134,788	151,077	12·09	171,006	13·19	184,914	8·13
Hereford	72,818	77,093	5·87	82,959	7·61	87,340	5·28
Hertford	75,717	86,060	13·53	99,337	15·56	108,368	9·09
Huntingdon	26,642	29,449	10·54	34,371	16·71	36,678	6·71
Kent	162,298	183,878	13·30	215,063	16·96	235,277	9·40
Lancaster	298,279	359,240	20·44	431,803	20·20	489,200	13·29
Leicester	93,694	104,505	11·54	117,048	12·00	123,601	5·60
Lincoln	156,043	174,440	11·79	206,500	18·38	229,137	10·96
Middlesex	62,515	74,326	18·89	88,031	18·44	99,834	13·41
Monmouth	35,771	49,672	38·86	59,702	20·19	77,937	30·54
Norfolk	189,770	203,430	7·20	235,076	15·56	260,571	10·84
Northampton	105,919	112,733	6·43	128,708	14·17	137,813	7·07
Northumberland	102,327	111,554	9·02	123,380	10·60	133,878	8·51
Nottingham	93,254	108,495	16·34	123,107	13·47	147,164	19·54
Oxford	79,866	85,257	6·75	96,487	13·17	105,456	9·29
Rutland	13,245	13,177	−0·51	14,697	11·53	15,238	3·68
Salop	92,914	102,831	10·67	108,859	5·86	115,795	6·37
Somerset	204,744	226,294	10·52	263,986	16·66	301,317	14·14
Southampton	138,366	152,161	9·97	171,756	12·88	186,620	8·65
Stafford	129,395	152,382	17·76	174,110	14·26	199,444	14·55
Suffolk	181,241	196,844	8·61	226,623	15·13	244,903	8·07
Surrey	80,094	91,311	14·00	105,383	15·41	118,789	12·72
Sussex	103,905	118,041	13·61	135,389	14·70	145,040	7·13
Warwick	98,686	104,543	3·52	120,340	15·11	125,711	4·46
Westmoreland	32,790	37,129	13·23	40,921	10·21	43,464	6·21
Wilts	97,255	101,372	4·23	112,806	11·28	121,497	7·71
Worcester	105,374	118,996	12·93	132,297	11·18	147,503	11·49
York, E., N., & W. R.	560,527	638,615	13·93	734,337	14·99	804,955	9·62
Total	4,829,219	5,418,470	12·20	6,215,969	14·72	6,878,176	10·65
WALES.							
Anglesey	27,523	29,998	8·99	35,699	19·00	37,773	5·81
Brecon	30,742	35,488	15·44	40,920	15·31	44,420	8·55
Cardigan	36,967	43,565	17·85	49,026	12·54	54,618	11·41
Carmarthen	59,205	66,730	12·71	77,242	15·75	84,605	9·53
Carnarvon	32,901	39,037	18·65	44,433	13·82	49,899	12·30
Denbigh	50,544	53,577	6·00	64,357	20·12	68,970	7·17
Flint	33,085	38,544	16·50	44,593	15·69	49,236	10·41
Glamorgan	43,288	49,643	14·68	55,999	12·80	64,390	14·98
Merioneth	29,506	30,854	4·57	34,382	11·43	35,315	2·71
Montgomery	38,829	40,428	4·12	44,649	10·44	47,301	5·94
Pembroke	48,186	51,546	6·97	60,457	17·29	65,542	8·41
Radnor	17,296	18,580	7·42	20,435	9·98	22,385	9·54
Total	448,072	497,990	11·14	572,192	14·90	624,454	9·13
Total England & Wales	5,277,291	5,916,460	12·11	6,788,161	14·74	7,502,630	10·52

or Decrease per Cent. of the Rural Districts between 1801 and 1871.

1841.	Increase	1851.	Increase	1861.	Increase	1871.	Increase	County.
No.	Per cnt.	No.	Per cnt.	No.	Per cnt.	No.	Per cnt.	
78,656	8·54	85,523	8·73	87,926	2·81	91,185	3·70	Bedford
98,837	8·21	103,544	4·76	104,552	0·97	111,203	6·36	Berks
98,123	7·12	102,875	4·84	107,283	4·28	111,833	4·24	Buckingham
114,259	13·89	127,630	11·70	122,773	−3·81	125,554	2·26	Cambridge
228,853	3·75	224,915	−1·72	242,688	7·90	282,454	16·38	Chester
243,979	10·27	247,048	1·26	254,036	2·83	243,806	−4·03	Cornwall
102,768	3·76	108,950	6·02	113,151	3·86	124,029	9·62	Cumberland
186,382	8·46	192,427	3·25	218,788	13·70	238,923	9·20	Derby
313,459	5·38	310,266	−1·02	297,061	−4·26	294,436	−0·88	Devon
116,692	10·26	121,204	3·86	124,383	2·62	128,617	3·41	Dorset
161,600	27·77	208,706	29·15	279,054	33·70	380,375	36·31	Durham
272,571	7·97	285,184	4·63	307,513	7·83	352,358	14·59	Essex
199,504	7·89	210,044	5·28	218,319	3·94	228,034	4·45	Gloucester
88,695	1·55	89,526	0·94	92,525	3·35	92,098	−0·47	Hereford
115,866	6·92	121,039	4·46	123,998	3·6	138,520	10·43	Hertford
40,580	10·64	44,253	9·05	44,342	0·20	43,557	−1·77	Huntingdon
250,309	6·39	257,087	2·71	276,461	7·54	309,615	11·99	Kent
550,409	12·51	620,174	12·67	747,064	20·46	858,082	14·86	Lancaster
131,301	6·23	134,517	2·45	134,363	−0·11	136,486	1·59	Leicester
259,378	13·20	284,390	9·64	284,119	−0·08	284,659	0·18	Lincoln
117,524	17·72	125,334	6·64	155,728	24·26	225,942	45·08	Middlesex
109,412	40·38	122,547	12·01	136,060	11·02	152,835	12·33	Monmouth
273,606	5·01	287,842	5·20	273,558	−4·96	263,576	−3·65	Norfolk
148,440	7·71	154,076	3·80	158,400	2·80	152,718	−3·59	Northampton
143,903	7·49	154,189	7·15	165,487	7·33	168,695	1·94	Northumberland
168,056	14·20	180,332	7·30	185,604	2·92	195,504	5·33	Nottingham
106,764	1·24	110,112	3·14	109,583	−0·48	110,768	1·08	Oxford
16,542	8·56	17,884	8·11	16,716	−6·53	16,383	−1·99	Rutland
122,200	5·53	120,622	−1·29	123,666	2·53	129,717	4·89	Salop
324,196	7·59	329,382	1·60	331,404	0·61	346,459	4·54	Somerset
205,823	10·29	219,205	6·51	254,436	16·07	275,939	8·45	Southampton
237,818	19·25	262,064	10·19	319,305	21·85	371,004	16·19	Stafford
256,172	4·60	268,105	4·66	263,869	−1·56	268,118	1·61	Suffolk
138,457	16·56	147,814	6·75	182,559	23·51	258,124	41·39	Surrey
155,671	7·33	161,877	3·99	161,280	−0·37	171,156	6·13	Sussex
139,841	11·24	147,882	5·75	163,791	10·76	184,781	12·82	Warwick
44,935	3·39	46,458	3·39	48,788	5·01	51,561	5·69	Westmoreland
129,304	6·42	128,795	−0·39	124,471	−3·36	122,568	−1·53	Wilts
156,506	6·11	173,955	11·15	196,946	13·22	175,646	−10·82	Worcester
893,382	10·98	947,191	6·02	1,020,385	7·73	1,095,016	7·31	York, E., N., & W. R.
7,540,773	9·63	7,984,968	5·89	8,572,465	7·36	9,312,337	8·63	**Total**
								WALES.
40,495	7·21	44,575	10·08	41,334	−7·27	37,368	−9·60	Anglesey
50,286	13·21	55,404	10·18	55,988	1·06	53,610	−4·24	Brecon
57,470	5·22	59,036	2·73	60,599	2·65	58,956	−2·71	Cardigan
88,273	4·34	89,471	1·36	88,502	−1·08	88,080	−0·48	Carmarthen
62,090	24·43	65,660	5·75	72,787	10·85	78,581	7·96	Carnarvon
73,028	5·88	75,969	4·03	82,890	9·11	84,879	2·40	Denbigh
52,331	6·29	53,647	2·52	55,176	2·85	57,354	3·95	Flint
82,336	27·87	103,655	25·89	141,290	36·31	183,214	29·67	Glamorgan
39,332	11·38	38,843	−1·24	38,963	0·31	46,598	19·59	Merioneth
49,907	5·51	48,434	−2·95	47,651	−1·62	48,143	1·03	Montgomery
70,094	6·94	71,634	2·20	68,199	−4·80	63,280	−7·21	Pembroke
22,980	2·66	22,371	−2·65	23,120	3·35	23,240	0·52	Radnor
688,622	10·28	728,699	5·82	776,499	6·56	823,303	6·03	**Total**
8,229,395	9·69	8,713,667	5·88	9,348,964	7·29	10,135,640	8·41	Total England & Wales

TABLE F.—*England and Wales.* *Population, and Increase or*

County.	1801.	1811.	Increase.	1821.	Increase.	1831.	Increase.
	No.	No.	Per cent.	No.	Per cent.	No.	Per cent.
Bedford	63,393	70,213	10·76	84,052	19·71	95,483	13·60
Berks	110,480	119,430	8·10	132,639	11·06	146,234	10·25
Buckingham	108,132	118,065	9·19	135,133	14·45	146,977	8·76
Cambridge	89,346	101,109	13·16	122,387	21·05	143,955	17·63
Chester	192,305	227,031	18·05	270,098	18·97	334,391	23·80
Cornwall	192,281	220,525	14·69	261,045	18·38	301,306	15·42
Cumberland	117,230	133,665	14·02	156,124	16·80	169,262	8·42
Derby	161,567	185,487	14·80	213,651	15·18	237,170	11·01
Devon	340,308	382,778	12·48	438,417	14·54	493,908	12·66
Dorset	114,452	124,718	8·97	144,930	16·21	159,385	9·98
Durham	149,384	165,293	10·65	193,511	17·07	239,256	23·64
Essex	227,682	252,473	10·89	289,424	14·64	317,507	9·71
Gloucester	250,723	285,955	14·06	336,190	17·56	387,398	15·23
Hereford	88,436	93,526	5·75	102,669	9·78	110,617	7·74
Hertford	97,393	111,225	14·23	129,731	16·61	142,844	10·11
Huntingdon	37,568	42,208	12·35	48,946	15·97	53,192	8·68
Kent	308,667	371,701	20·42	427,224	14·94	479,558	12·25
Lancaster	673,486	828,499	23·02	1,052,948	27·09	1,336,854	26·97
Leicester	130,082	150,559	15·74	174,571	15·95	197,003	12·85
Lincoln	208,625	237,634	13·90	283,058	19·12	317,465	12·16
Middlesex	818,129	953,774	16·58	1,145,057	20·06	1,358,330	18·62
Monmouth	45,568	62,105	36·29	75,801	22·05	98,126	29·45
Norfolk	273,479	291,947	6·75	344,368	17·95	390,054	13·27
Northampton	131,525	141,353	7·47	163,097	15·39	179,336	9·96
Northumberland	168,078	183,269	9·04	212,589	16·00	236,959	11·47
Nottingham	140,350	162,964	16·11	186,873	14·67	225,327	20·58
Oxford	111,977	120,376	7·50	138,224	14·82	153,526	11·08
Rutland	16,300	16,380	0·49	18,487	12·87	19,385	4·86
Salop	169,248	184,973	9·29	198,311	7·21	213,518	7·67
Somerset	273,577	302,836	10·70	355,789	17·48	403,795	13·49
Southampton	219,290	246,514	12·41	282,897	14·76	313,976	10·99
Stafford	242,693	294,540	21·36	345,972	17·46	409,480	18·36
Suffolk	214,404	233,963	9·12	271,541	16·06	296,317	9·12
Surrey	268,233	323,851	20·73	399,441	23·34	486,434	21·78
Sussex	159,471	190,343	19·36	233,328	22·59	272,644	16·85
Warwick	206,798	228,906	10·69	274,482	19·91	336,645	22·65
Westmoreland	40,805	45,922	12·54	51,359	11·84	55,041	7·17
Wilts	183,820	191,853	4·37	219,574	14·45	237,244	8·05
Worcester	146,441	168,982	15·39	194,074	14·85	222,655	14·73
York, E.,N., and W. R.	859,133	986,076	14·78	1,173,895	19·05	1,371,966	16·87
Total	8,350,859	9,553,021	14·40	11,281,883	18·10	13,090,523	16·04
WALES.							
Anglesey	33,806	37,045	9·58	45,063	21·64	48,325	7·24
Brecon	32,325	37,735	16·74	43,826	16·14	47,763	8·98
Cardigan	42,956	50,260	17·00	57,784	14·97	64,780	12·11
Carmarthen	67,317	77,217	14·71	90,239	16·86	100,740	11·64
Carnarvon	41,521	49,655	19·59	58,099	17·01	66,818	15·01
Denbigh	60,299	64,249	6·55	76,428	18·95	82,665	8·16
Flint	39,469	45,937	16·39	53,893	17·32	60,244	11·78
Glamorgan	70,879	85,067	20·02	102,073	19·99	126,612	24·04
Merioneth	29,506	30,854	4·57	34,382	11·43	35,315	2·71
Montgomery	48,184	52,184	8·30	60,245	15·45	66,844	10·95
Pembroke	56,280	60,615	7·70	73,788	21·73	81,425	10·35
Radnor	19,135	20,417	6·70	22,533	10·36	24,743	9·81
Total	541,677	611,235	12·84	718,353	17·52	806,274	12·24
Total England & Wales	8,892,536	10,164,256	14·30	12,000,236	18·06	13,896,797	15·81

Decrease per Cent. of each County for each Decade between 1801 and 1871.

1841.	Increase	1851.	Increase	1861.	Increase	1871.	Increase	County.
No.	Per cnt.	No.	Per cnt.	No.	Per cnt.	No.	Per cnt.	
107,936	13·04	124,478	15·33	135,287	8·68	146,257	8·11	Bedford
161,759	10·62	170,065	5·14	176,256	3·64	196,475	11·47	Berks
156,439	6·44	163,723	4·66	167,993	2·61	175,879	4·69	Buckingham
164,459	14·24	185,405	12·74	176,016	− 5·06	186,906	6·19	Cambridge
395,660	18·32	455,725	15·18	505,428	10·90	561,201	11·04	Chester
342,159	13·56	355,558	3·92	369,390	3·89	362,343	− 1·91	Cornwall
178,038	5·19	195,492	9·80	205,276	5·01	220,253	7·29	Cumberland
272,202	14·77	296,084	8·77	339,327	14·61	379,394	11·80	Derby
532,959	7·91	567,098	6·41	584,373	3·05	601,374	2·91	Devon
175,054	9·82	184,207	5·23	188,789	2·49	195,537	3·59	Dorset
307,963	28·71	390,997	26·96	508,666	30·09	685,089	34·68	Durham
344,979	8·65	369,318	7·06	404,851	9·62	466,436	15·21	Essex
431,495	11·38	458,805	6·33	485,770	5·88	534,640	10·06	Gloucester
113,272	2·40	115,489	1·96	123,712	7·12	125,370	1·33	Hereford
156,660	9·68	167,298	6·79	173,280	3·57	192,226	10·94	Hertford
58,549	10·07	64,183	9·62	64,250	0·10	63,708	− 0·84	Huntingdon
549,353	14·55	615,766	12·09	733,887	19·18	848,294	15·59	Kent
1,667,054	24·70	2,031,236	21·84	2,429,440	19·61	2,819,495	16·06	Lancaster
215,867	9·58	230,308	6·69	237,412	3·08	269,311	13·44	Leicester
362,602	14·22	407,222	12·31	412,246	1·23	436,599	5·91	Lincoln
1,576,636	16·07	1,886,576	19·66	2,206,485	16·96	2,539,765	15·11	Middlesex
134,368	36·93	157,418	17·15	174,633	10·94	195,448	11·92	Monmouth
412,664	5·80	442,714	7·28	434,798	− 1·79	438,656	0·89	Norfolk
199,228	11·09	212,380	6·60	227,704	7·21	243,891	7·11	Northampton
266,020	12·26	303,568	14·12	343,025	13·00	386,646	12·72	Northumberland
249,910	10·91	270,427	8·21	293,867	8·67	319,758	8·81	Nottingham
163,127	6·25	170,439	4·48	170,944	0·29	177,975	4·12	Oxford
21,302	9·89	22,983	7·89	21,861	− 4·88	22,073	0·97	Rutland
225,820	5·76	229,341	1·56	240,959	5·07	248,111	2·97	Salop
435,599	7·87	443,916	1·91	444,873	0·21	463,483	4·18	Somerset
354,682	12·96	405,370	14·29	481,815	18·86	544,684	13·05	Southampton
509,472	24·42	608,716	19·48	746,943	22·71	858,326	14·91	Stafford
315,073	6·32	337,215	7·02	337,070	− 0·04	348,869	3·50	Suffolk
584,036	20·07	683,082	16·96	831,093	21·67	1,091,635	31·35	Surrey
300,075	10·07	336,844	12·25	363,735	7·99	417,456	14·77	Sussex
401,703	19·32	475,013	18·25	561,855	18·29	634,189	12·87	Warwick
56,454	2·57	58,287	3·25	60,817	4·34	65,010	6·90	Westmoreland
256,280	8·03	254,221	− 0·80	249,311	− 1·93	257,177	3·16	Wilts
248,460	11·59	276,926	11·46	307,397	11·00	338,837	10·23	Worcester
1,592,059	16·04	1,797,995	12·93	2,033,610	13·10	2,436,355	19·81	York, E., N., & W. R.
14,997,427	14·56	16,921,888	12·84	18,954,444	12·01	21,495,131	13·41	Total
								WALES.
50,891	5·31	57,327	12·65	54,609	− 4·74	51,040	− 6·54	Anglesey
55,603	16·42	61,474	10·56	61,627	0·25	59,901	− 2·80	Brecon
68,766	6·15	70,796	2·95	72,245	2·05	73,441	1·66	Cardigan
106,326	5·55	110,632	4·05	111,796	1·05	115,710	3·50	Carmarthen
81,093	21·36	87,870	8·36	95,694	8·90	106,121	10·90	Carnarvon
88,478	7·03	92,583	4·64	100,778	8·85	105,102	4·29	Denbigh
66,919	11·08	68,156	1·85	69,737	2·32	76,312	9·43	Flint
171,188	35·21	231,849	35·43	317,752	37·05	397,859	25·21	Glamorgan
39,332	11·38	38,843	− 1·24	38,963	0·31	46,598	19·59	Merioneth
69,607	4·13	67,335	− 3·26	66,919	− 0·62	67,623	1·05	Montgomery
88,044	8·13	94,140	6·92	96,278	2·27	91,998	− 4·45	Pembroke
25,458	2·89	24,716	− 2·91	25,382	2·70	25,430	0·19	Radnor
911,705	13·08	1,005,721	10·31	1,111,780	10·55	1,217,135	9·47	Total
15,909,132	14·48	17,927,609	12·69	20,066,224	11·93	22,712,266	13·19	Total England & Wales

D

TABLE G.—*Population and Rates of Increase or Decrease per Cent. of each Town in*
(The Towns arranged in the Numerical

		Boundary Taken.	1801.	1811.	Rate of Increase.	1821.	Rate of Increase.
1	London	Reg. General's tabl·s of mortality	958,863	1,138,815	18·77	1,378,947	21·09
2	Liverpool (Lancashire)	M. and P.	82,295	104,104	26·50	138,354	32·90
3	Manchester (Lancashire)	P.	76,788	91,130	18·68	129,035	41·60
4	Birmingham (Warwickshire)	M. and P.	70,670	82,753	17·10	101,722	22·92
5	Leeds (Yorkshire)	,,	53,162	62,534	17·63	83,796	34·00
6	Sheffield (Yorkshire)	,,	45,755	53,231	16·34	65,275	22·63
7	Bristol (Gloucestershire)	,,	61,153	71,433	16·81	85,108	19·14
8	Wolverhampton (Staffordshire)	P.	30,584	43,190	*41·22	53,011	22·74
9	Bradford (Yorkshire)	M. and P.	13,264	16,012	20·73	26,307	64·30
10	Newcastle-on-Tyne (Northumberland)	,,	33,048	32,573	− 1·44	41,794	28·31
11	Stoke-upon-Trent (Staffordshire)	P. (e)	23,278	31,557	*35·57	40,237	27·51
12	Hull (Yorkshire)	,,	29,580	37,005	25·10	44,520	20·31
13	Salford (Lancashire)	M. and P.	18,088	24,744	36·80	32,600	31·75
14	Portsmouth (Hants)	,,	33,226	41,587	*25·16	46,743	12·40
15	Oldham (Lancashire)	P. (e)	21,766	29,479	*35·43	38,201	29·59
16	Sunderland (Durham)	,,	24,998	25,821	3·29	31,891	23·51
17	Brighton (Sussex)	,,	7,440	12,205	64·05	24,741	*102·71
18	Merthyr Tydvil (Glamorganshire)	,,	10,127	14,945	47·58	20,959	40·24
19	Leicester (Leicestershire)	M. and P.	17,005	23,453	37·92	31,036	32·33
20	Bolton (Lancashire)	P. (e)	17,966	24,799	*38·03	32,045	29·22
21	Nottingham (Nottinghamshire)	M. and P.	28,801	34,030	18·15	40,190	18·10
22	Preston (Lancashire)	M. & P. (e)	12,174	17,360	42·50	24,859	43·20
23	Blackburn (Lancashire)	P. (e)	11,980	15,083	25·90	21,940	*45·46
24	Dudley (Worcestershire)	,,	10,107	13,925	37·78	18,211	30·78
25	Norwich (Norfolk)	M. and P.	36,854	36,256	− 1·62	50,288	*38·70
26	Huddersfield (Yorkshire)	P. (e)	7,268	9,671	33·06	13,284	37·36
27	Plymouth (Devonshire)	,,	16,040	20,803	29·70	21,591	3·79
28	Birkenhead (Cheshire)	,,	667	795	19·19	1,313	65·15
29	Halifax (Yorkshire)	,,	12,010	12,766	6·29	17,056	33·61
30	Devonport (Devonshire)	P.	27,154	35,257	*29·84	39,621	12·38
31	Rochdale (Lancashire)	P. (e)	8,542	10,753	25·88	14,017	30·36
32	Derby (Derbyshire)	,,	10,832	13,043	20·41	17,423	33·58
33	Southampton (Hants)	M. and P.	7,913	9,617	21·53	13,353	38·85
34	Bath (Somersetshire)	P. (e)	33,196	34,408	15·70	46,700	*21·59
35	Stockport (Cheshire)	M. and P.	14,830	17,554	18·31	21,726	23·83
36	Swansea (Glamorganshire)	M.	10,117	11,963	18·25	14,896	24·52
37	Gateshead (Durham)	M. and P.	8,597	8,782	2·15	11,767	33·99
38	Middlesborough (Yorkshire)	—	239	212	−11·30	236	11·32
39	Walsall (Staffordshire)	P. (e)	10,399	11,189	7·60	11,914	6·48
40	Chatham (Kent)	,,	12,940	15,787	22·00	19,177	21·47
41	Southport (Lancashire)	M.	3,201	3,999	24·93	4,614	15·38
42	Northampton (Northamptonshire)	P. (e)	7,020	8,427	20·04	10,793	28·07
43	South Shields (Durham)	M. and P.	11,011	15,165	*37·72	16,503	8·82
44	Cheltenham (Gloucestershire)	P. (e)	3,076	8,325	*170·65	13,396	60·91
45	Exeter city (Devonshire)	,,	17,412	18,896	8·52	23,479	24·45
46	York city (Yorkshire)	M.	16,846	19,099	13·37	21,711	13·68
47	Cricklade (Wiltshire)	P.	22,139	23,777	7·40	27,038	13·67
48	Ipswich (Suffolk)	M. and P.	11,277	13,670	21·12	17,186	25·72
49	Yarmouth (Norfolk)	M.	16,573	20,448	*23·38	21,007	2·73
50	Coventry (Warwickshire)	P. (e)	16,034	17,923	11·78	21,448	19·67
51	Bury (Lancashire)	P.	9,152	11,302	23·49	13,480	19·27

...gland and **Wales** *of Twenty Thousand Inhabitants and upwards,* 1801 *and* 1871.
...der according to the Census of 1871.)

1831.	Rate of Increase.	1841.	Rate of Increase.	1851.	Rate of Increase.	1861.	Rate of Increase.	1871.	Rate of Increase.	
354,994	20·02	1,948,417	17·73	2,632,236	*21·24	2,803,989	18·70	3,254,260	16·06	1
01,751	*45·82	286,487	42·00	375,955	31·23	443,938	18·08	493,405	11·14	2
87,022	*44·93	242,983	29·92	316,213	30·14	357,979	13·21	379,374	5·97	3
43,986	*41·55	182,922	27·04	232,841	27·29	296,076	27·16	343,787	16·11	4
23,393	*47·25	152,074	23·24	172,270	13·28	207,165	20·26	259,212	25·12	5
91,692	*40·47	111,091	21·16	135,310	21·80	185,172	36·85	239,946	29·58	6
04,408	*22·68	125,146	19·87	137,328	9·73	154,093	12·20	182,552	18·47	7
67,514	27·36	93,245	38·11	119,748	28·42	147,670	23·32	156,978	6·30	8
43,527	*65·46	66,715	53·27	103,778	55·55	106,218	2·35	145,830	37·29	9
53,613	28·29	70,337	*31·19	87,784	24·81	109,108	24·29	128,443	17·72	10
51,589	28·21	68,444	32·67	84,027	22·77	101,207	20·45	124,493	23·00	11
51,911	16·60	67,308	*29·66	84,690	25·82	97,661	15·32	123,408	26·37	12
50,810	*55·86	68,386	34·59	85,108	24·45	102,249	20·38	121,401	18·50	13
50,389	7·80	53,032	5·24	72,096	*35·95	94,799	31·49	113,569	19·80	14
50,513	32·23	60,451	19·67	72,357	19·70	94,344	30·39	113,100	19·88	15
40,735	27·73	53,335	*30·93	67,391	26·36	85,797	27·31	104,490	21·79	16
41,994	69·73	49,170	17·09	69,673	41·70	87,317	25·33	103,758	18·83	17
27,281	30·16	43,031	*57·73	63,080	46·59	83,875	32·97	96,891	15·52	18
40,639	30·94	50,806	25·02	60,584	19·25	68,056	12·33	95,220	*39·91	19
42,245	31·83	51,029	20·79	61,171	19·88	70,395	15·08	92,658	31·63	20
50,220	24·96	52,164	3·87	57,407	10·05	74,693	*30·11	86,621	15·97	21
33,871	36·25	50,887	*50·24	69,542	36·66	82,985	19·33	85,427	2·94	22
27,091	23·48	36,629	35·21	46,536	27·05	63,126	35·65	82,853	31·25	23
23,430	28·66	31,232	33·30	37,962	21·55	44,975	18·47	82,249	*82·88	24
61,116	21·53	62,344	2·01	68,713	10·22	74,891	8·99	80,386	7·34	25
19,035	43·29	25,068	31·70	30,880	23·19	34,877	12·94	74,358	*113·20	26
31,080	*43·95	36,520	17·50	52,221	42·99	62,599	19·87	70,091	11·97	27
4,195	*219·50	11,563	175·64	34,469	198·10	51,649	49·84	65,971	27·73	28
21,552	23·48	27,520	27·69	33,582	22·03	37,014	10·22	65,500	*76·96	29
44,454	12·20	43,532	−2·07	50,159	15·22	64,783	29·15	64,034	−1·16	30
19,041	35·84	24,272	27·47	29,195	20·28	38,184	30·79	63,485	*66·26	31
23,627	35·61	32,741	38·57	40,609	24·03	43,091	6·11	61,381	*42·45	32
19,324	*44·72	27,744	43·57	35,305	27·25	46,960	33·01	53,741	14·44	33
50,800	8·78	53,196	4·72	54,240	1·96	55,528	−3·16	53,714	2·26	34
25,469	17·23	50,154	*96·92	53,835	7·34	54,861	1·91	53,014	−3·37	35
19,672	32·06	24,604	25·07	31,461	27·87	41,606	*32·25	51,720	24·31	36
15,177	28·98	20,123	32·59	25,568	27·06	33,587	31·36	48,592	*44·67	37
383	62·29	5,709	*1390·60	7,893	38·25	18,992	140·62	46,621	145·48	38
15,066	26·46	19,857	31·80	25,680	29·33	37,760	*47·04	46,098	22·08	39
21,124	10·15	24,269	14·89	28,424	17·12	36,177	*27·27	45,792	26·58	40
6,101	32·23	8,994	47·42	14,866	65·28	18,396	23·75	45,124	*145·29	41
15,351	*42·24	21,242	38·38	26,657	25·49	32,813	23·10	45,080	37·39	42
18,756	13·65	23,072	23·01	28,974	25·58	35,239	21·62	44,722	26·91	43
22,942	71·26	31,411	36·91	35,501	11·59	39,663	13·24	44,519	12·16	44
28,242	20·29	37,231	*31·83	40,688	9·29	41,749	2·61	44,226	5·93	45
26,260	20·95	28,842	9·83	36,303	*25·87	40,433	11·38	43,796	8·32	46
59,118	7·73	55,409	*21·61	35,543	0·27	36,893	3·91	43,622	18·24	47
20,201	17·54	25,384	25·66	32,914	*29·66	37,950	15·30	42,947	13·17	48
24,535	16·80	27,865	13·57	30,879	10·82	34,810	12·73	41,819	20·14	49
27,298	*27·27	31,032	13·68	36,812	18·63	41,647	13·13	41,348	−0·72	50
19,140	*41·99	24,846	29·81	31,262	25·82	37,563	20·15	41,344	10·07	51

TABLE G.—

		Boundary Taken.	1801.	1811.	Rate of Increase.	1821.	Rate of Increase.
52	Burnley (Lancashire)	P. (*new*)	3,918	5,405	37·96	8,242	*52·49
53	Hanley (Staffordshire)	M.	4,338	5,345	23·21	6,585	23·20
54	Cardiff (Glamorganshire)	,,	1,870	2,457	31·39	3,521	43·30
55	Wigan (Lancashire)	M. and P.	10,989	14,060	*27·95	17,716	26·00
56 {	Tynemouth and North Shields (Northumberland) }	,,	13,171	17,548	*33·23	23,173	32·05
57	Strood (Gloucestershire)	P.	27,732	28,448	2·58	36,340	*27·74
58	Worcester city (Worcestershire)	P (*e*)	11,460	13,814	20·54	17,023	23·23
59	New Shoreham (Sussex)	P.	16,104	18,690	16·06	22,722	*21·57
60 {	Ashton-under-Lyne (Lancashire) }	P. (*e*)	6,391	7,959	24·53	9,222	15·87
61	Macclesfield (Cheshire)	,,	8,743	12,299	40·67	17,746	*44·29
62	Chester (Cheshire)	,,	15,174	16,140	6·37	19,949	*23·60
63	Cambridge (Cambridgeshire)	,,	10,087	11,108	10·12	14,142	27·31
64	Hastings (Sussex)	,,	2,982	3,848	29·04	6,111	58·81
65	Warrington (Lancashire)	P.	11,321	12,682	12·02	14,822	16·88
66	Reading (Berks)	M. (*e*)	9,742	10,788	10·74	12,867	19·27
67	Oxford (Oxfordshire)	P. (*e*)	11,694	12,931	10·58	16,364	*26·55
68	Carlisle (Cumberland)	M. and P.	9,415	11,476	21·89	14,416	25·62
69	Morpeth (Northumberland)	M.	4,859	5,288	8·83	5,853	10·69
70	Aylesbury (Buckinghamshire)	P.	16,993	18,435	8·49	21,717	*17·80
71	Dover (Kent)	M. and P.	8,028	11,230	*39·88	12,661	12·77
72	Wakefield (Yorkshire)	P (*e*)	10,581	11,393	7·67	14,164	*24·32
73	Stockton (Durham)	—	3,935	4,187	6·38	2,956	18·36
74	Darlington (Durham)	M.	4,670	5,059	8·33	5,750	13·66
75 {	Newport Mon. (Monmouthshire) }	,,	1,423	3,025	*112·58	4,951	63·67
76	Great Grimsby (Lincolnshire)	,,	3,143	4,542	44·51	5,183	14·11
77	Lincoln (Lincolnshire)	M. and P.	7,197	8,599	19·48	9,995	*16·23
78	Colchester (Essex)	,,	11,520	12,544	8·89	14,016	11·73
79	Maidstone (Kent)	P.	8,027	9,443	17·64	12,508	32·46
80	Wednesbury (Staffordshire)	M.	4,160	5,372	29·14	6,471	20·46
81	Dewsbury (Yorkshire)	,,	4,566	5,059	10·80	6,380	26·11
82	Keighley (Yorkshire)	,,	5,745	6,864	19·48	9,223	34·37
83	Scarborough (Yorkshire)	P. and M.	6,688	7,067	5·67	8,533	20·75
84	Hythe (Kent)	M.	2,987	4,768	59·62	4,489	-5·85
85	Shrewsbury (Salop)	M. and P.	14,739	16,825	14·15	19,854	*18·00
86	Heywood (Lancashire)	M.	6,697	8,001	19·47	9,922	24·01
87	Stratford (Essex)	,,	3,910	4,905	15·45	5,882	19·92
88	Barnsley (Yorkshire)	,,	3,606	5,104	41·54	8,284	*62·31
89	Torquay (Devonshire)	,,	838	1,350	61·10	1,925	42·59
90	Over Darwen (Lancashire)	,,	3,587	4,411	22·97	6,711	*52·14
91	Gravesend (Kent)	,,	4,539	5,589	23·13	6,583	17·79
92	Wenlock (Salop)	P.	16,304	16,805	3·07	17,265	2·74
93	Staleybridge (Cheshire)	M.		(Not ascertainable			
94	Canterbury city (Kent)	M. and P.	9,000	10,200	13·33	12,779	25·28
95	Leamington (Warwickshire)	M.	315	543	72·38	2,183	*302·02
96	Batley (Yorkshire)	,,	2,574	2,957	14·88	3,717	25·70
97	Kidderminster (Worcestershire)	,,	6,803	8,753	28·66	11,444	30·74
98	Luton (Bedfordshire)	,,	3,095	3,716	20·07	4,529	21·88
99	Brentford (Middlesex)	,,	5,035	5,361	6·48	6,608	23·26
	Total	—	2,404,153	2,878,039	19·71	3,582,029	24·46

Note.—Where marked (*e*) the boundary
        * The decades thus marked show when

Contd.

1831.	Rate of Increase.	1841.	Rate of Increase.	1851.	Rate of Increase	1861.	Rate of Incre-se.	1871.	Rate of Increase.	
10,026	21·65	14,224	41·87	20,828	46·43	28,700	37·79	40,858	42·36	52
8,282	25·77	10,218	23·38	25,369	*148·27	31,953	25·95	39,976	25·11	53
6,187	75·72	10,077	62·87	18,351	*82·11	32,954	79·58	39,675	20·40	54
20,774	17·26	25,517	22·83	31,941	25·18	37,658	17·90	39,110	3·86	55
23,206	0·14	25,416	9·52	29,170	14·77	34,021	16·63	38,941	14·46	56
59,932	9·89	37,992	-4·86	36,535	-3·84	35,517	-2·79	38,610	8·71	57
18,610	9·32	27,004	*45·10	27,528	1·94	31,227	13·44	38,116	22·06	58
25,356	11·59	28,638	12·94	30,553	6·69	32,622	6·77	37,984	16·44	59
14,035	52·19	22,678	*61·58	29,791	31·37	33,917	13·85	37,389	10·24	60
23,129	30·33	32,629	41·07	39,048	19·67	36,101	-7·55	35,450	-1·80	61
21,344	7·00	23,866	11·82	27,766	16·34	31,110	12·04	35,257	13·33	62
20,917	47·91	24,453	16·90	27,815	13·75	26,361	-5·23	33,996	28·96	63
10,097	*65·23	11,617	15·05	17,011	46·43	22,910	34·68	33,337	45·51	64
18,184	*22·68	21,346	17·39	23,363	9·45	26,947	15·34	33,050	22·65	65
15,595	21·20	18,939	21·43	21,456	13·30	25,045	16·75	32,313	*29·02	66
20,649	*26·19	24,258	17·48	27,843	14·78	27,560	-1·02	31,404	13·95	67
18,865	30·86	21,550	14·23	26,310	22·09	29,417	11·81	31,049	5·55	68
6,577	12·37	7,160	8·86	10,012	39·83	13,794	37·76	30,239	*119·22	69
24,162	11·26	25,337	4·86	26,794	5·75	27,090	1·10	28,760	6·16	70
15,645	23·54	19,189	22·65	22,244	15·92	25,325	13·85	28,506	12·56	71
15,932	12·48	18,842	18·27	22,057	17·06	23,150	4·95	28,069	21·25	72
7,685	55·06	9,727	26·57	9,710	-0·17	13,357	37·56	27,738	*107·67	73
8,574	49·11	11,033	28·68	11,582	4·98	15,789	36·32	27,729	*75·62	74
7,062	42·64	10,815	53·15	19,323	78·68	23,249	20·32	27,069	16·43	75
6,589	27·13	6,698	1·65	12,263	*83·08	15,060	22·81	26,982	79·16	76
11,217	12·25	13,896	23·88	17,533	26·17	20,999	19·77	26,766	*27·46	77
16,167	15·35	17,790	10·04	19,443	9·29	23,809	*22·46	26,343	10·64	78
15,790	26·24	18,086	14·54	20,801	15·01	23,058	10·85	26,237	13·79	79
8,437	30·38	11,625	37·79	14,281	22·85	21,968	*53·83	25,030	13·94	80
8,272	29·66	10,600	28·14	14,049	32·54	18,148	29·18	24,764	*36·46	81
11,176	21·17	13,413	20·02	18,259	*36·13	18,819	3·07	24,704	31·37	82
8,760	2·66	10,060	14·84	12,915	28·38	18,377	*42·29	24,259	32·01	83
4,623	2·99	8,939	93·36	13,164	47·27	21,367	*62·31	24,078	12·69	84
21,297	7·27	21,518	1·04	23,104	7·37	25,784	11·60	23,406	-9·22	85
14,229	*43·41	18,720	31·56	19,872	6·15	22,349	12·47	23,394	4·68	86
6,991	18·85	7,690	9·99	10,586	37·66	15,994	*51·09	23,286	45·59	87
10,330	24·70	12,310	19·17	13,437	9·15	17,890	33·14	23,021	28·68	88
3,582	86·08	5,982	67·00	11,474	*91·81	16,419	43·10	21,657	31·90	89
6,972	3·89	9,348	34·08	11,702	25·18	16,492	40·93	21,278	29·02	90
9,445	43·28	15,670	*65·91	16,633	6·15	18,782	12·92	21,265	13·22	91
17,435	0·98	18,016	3·33	20,588	*14·27	21,590	4·87	21,208	-1·77	92
				20,760	—	24,921	*20·04	21,092	-15·36	93
13,679	7·04	17,904	*30·89	18,398	2·76	21,324	15·90	20,962	-1·70	94
6,209	184·43	12,864	107·18	15,724	22·23	17,402	10·67	20,910	20·16	95
4,841	30·24	7,076	46·17	9,308	31·54	14,873	*59·79	20,871	40·33	96
16,036	*40·13	15,427	-3·80	18,462	19·67	15,399	-16·59	20,814	35·16	97
5,693	25·70	7,748	36·10	12,787	*65·04	17,821	39·37	20,733	16·34	98
7,783	17·78	8,407	8·02	9,828	16·90	13,958	42·02	20,232	*44·95	99
4,520,055	26·19	5,572,175	23·28	6,885,001	23·56	8,218,209	19·36	9,800,887	19·25	

has been extended in 1871.
the maximum increase was reached.

Table H.—*United Kingdom. Enumerated Population of the United Kingdom and of its Constituent Parts, at each of the Censuses 1801 to 1871, with the Numbers of the Army, Navy, and Merchant Seamen belonging to the Kingdom.*

[Census Returns. Copy of Table 3, p. 4, of "General Report," vol. iv, for 1871.]

Census Years.	United Kingdom, including Islands in British Seas, and Army, Navy, and Merchant Seamen Abroad.	United Kingdom, including Islands in British Seas, but excluding Army, Navy. and Merchant Seamen Abroad.	United Kingdom, excluding Islands in British Seas, and Army, Navy, and Merchant Seamen Abroad.
1801........	16,237,300	15,795,287	15,717,287
'11........	18,509,116	18,006,580	17,926,580
'21........	21,272,187	20,982,092	20,893,584
'31........	24,392,485	24,132,294	24,028,584
'41........	27,057,923	26,854,969	26,730,929
'51........	27,745,949	27,533,755	27,390,629
'61........	29,321,288	29,070,932	28,927,485
'71........	31,845,379	31,629,299	31,484,661

Census Years.	England and Wales.	Scotland.	Ireland.	Islands in the British Seas.	Army, Navy, Marines, and Merchant Seamen belonging to the Kingdom.
1801........	8,892,536	1,608,420	5,216,331	78,000	442,013*
'11........	10,164,256	1,805,864	5,956,460	80,000	502,536*
'21........	12,000,236	2,091,521	6,801,827	89,508	289,095*
'31........	13,896,797	2,364,386	7,767,401	103,710	260,191*
'41........	15,914,148	2,620,184	8,196,597	124,040	202,954†
'51........	17,927,609	2,888,742	6,574,278	143,126	212,194‡
'61........	20,066,224	3,062,294	5,798,967	143,447	250,356‡
'71........	22,712,266	3,360,018	5,412,377	144,638	216'080‡

* At home and abroad.
† Abroad or on board vessels in ports, the latter being estimated at 28,520.
‡ Abroad only.

Table I.—*United Kingdom. Enumerated Population of the United Kingdom and of its Constituent Parts, including the Army, Navy, Marines, and Merchant Seamen Abroad, belonging to the Kingdom at each of the Censuses 1801 to 1871.*

[Copy of Table IV from "General Report," vol. iv, England and Wales.]

Census Years.	Enumerated Population.					Increase of Population.				
	United Kingdom.	England and Wales.	Scotland.	Ireland.	Islands in the British Seas.	United Kingdom.	England and Wales.	Scotland.	Ireland.	Islands in the British Seas.
1801 ...	16,237,300	9,156,171	1,678,452	5,319,867	82,810					
'11 ...	18,509,116	10,454,529	1,884,044	6,084,996	85,547	2,271,816	1,298,358	205,592	765,129	2,737
'21 ...	21,272,187	12,172,664	2,137,325	6,869,544	92,654	2,763,071	1,718,135	253,281	784,548	7,107
'31 ...	24,392,485	14,051,986	2,405,610	7,828,347	106,542	3,120,298	1,879,322	268,285	958,803	13,888
'41 ...	27,057,923	16,035,198	2,652,339	8,244,137	126,249	2,665,438	1,938,212	246,729	415,790 Decrease	19,707
'51 ...	27,745,949	18,054,170	2,922,362	6,623,982	145,435	688,026	2,018,972	270,023	1,620,155 Decrease	19,186
'61 ...	29,321,288	20,228,497	3,096,808	5,850,309	145,674	1,575,339	2,174,327	174,446	773,673 Decrease	239
'71 ...	31,845,379	22,856,164	3,392,559	5,449,186	147,470	2,524,091	2,627,667	295,751	401,123	1,796
Total increase, 1801 to 1871						15,608,079	13,699,993	1,714,107	129,319	64,660

Note.—The population of Ireland is estimated for the years 1801 and 1811.

TABLE K.—*Total Estimated Consumption of Coal in the United Kingdom on the Basis of Diminishing Ratios (Decreasing Rate of Increase* 5·831 *per Cent. per Decade).*

[Copy of Table III in Coal Commissioners' Report, p. xvi.]

Year.	Estimated Population of Great Britain. [000's omitted.]	Estimated Consumption of Coal per Head.	Total Estimated Home Consumption per Annum. [000's omitted.]	Total Consumption per Century, including Exportation.
	No.	Tons.	Tons.	
1871	26,063,	3·9636	103,300,	
1881	28,943,	4·4266	128,100,	Mln. Tons.
'91	31,955,	4·5786	146,300,	20,144 home consumption
1901	35,087,	4·6286	162,400,	1,200 exported
'11	38,326,	4·6446	178,000,	
'21	41,561,	4·6946	193,200,	21,344 total in century
'31	44,859,	4·6516	208,700,	
'41	48,316,	4·6526	224,800,	
'51	51,823,	4·6526	241,100,	
'61	55,365,	4·6526	257,600,	
'71 ...	58,928,	4·6526	274,200,	
1981	62,500,	4·6526	290,800,	
'91	66,070,	4·6526	307,400,	
2001	69,620,	4·6526	323,900,	
'11	73,140,	4·6526	340,300,	36,306 home consumption
'21	76,450,	4·6526	355,700,	1,200 exported
'31	79,880,	4·6526	371,700,	
'41	83,260,	4·6526	387,400,	37,506 total in century
'51	86,580,	4·6526	402,800,	
'61	89,820,	4·6526	417,900,	
'71	93,000,	4·6526	432,700,	
2081	96,080,	4·6526	447,000,	
'91	99,090,	4·6526	461,000,	
2101	102,010,	4·6526	474,600,	
'11 ...	104,850,	4·6526	487,800,	50,501 home consumption
'21	107,580,	4·6526	500,600,	1,200 exported
'31	110,230,	4·6526	512,900,	
'41	112,790,	4·6526	524,800,	51,701 total in century
'51	115,250,	4·6526	536,200,	
'61	117,620,	4·6526	547,300,	
'71	119,900,	4·6526	557,900,	
2181	122,180,	4·6526	568,000,	
'91	124,180,	4·6526	577,800,	35,465 home consumption
2201	126,200,	4·6526	587,200,	720 exported
'11	128,110,	4·6526	596,100,	
'21	129,950,	4·6526	604,600,	36,185 total for 60 years
'31	131,700,	4·6526	612,800,	

Total consumption in 360 years .... 146,736 millions

Table KA.—*Total Estimated Consumption of Coal in the United Kingdom on the Basis of Diminishing Rates of Increase of the Population (viz., 4·694 per Cent. Decrease per Decade).*

[Amended copy of Table III in Coal Commissioners' Report, p. xvi.]

Year.	Estimated Population of Great Britain. [000's omitted.]	Estimated Consumption of Coal per Head.	Total Estimated Home Consumption per Annum. [000's omitted.]	Total Consumption per Century, including Exportation.
	No.	Tons.	Tons.	
1871	26,249,	3·9636	104,039,	
1881	29,383,	4·4266	130,067, ⎫	
'91	32,727,	4·5786	149,844,	
1901	36,278,	4·6286	167,916,	Mln. Tons.
'11	40,029,	4·6446	185,919,	21,650 home consumption
'21	43,975,	4·6496	204,466,	1,200 exported
'31	48,105,	4·6516	223,765, ⎬	—————
'41	52,411,	4·6526	243,847,	22,850 total in century
'51	56,881,	4·6526	264,645,	—————
'61	61,505,	4·6526	286,158,	
'71	66,272,	4·6526	308,337, ⎭	
1981	71,170,	4·6526	331,126, ⎫	
'91	76,180,	4·6526	354,435,	
2001	81,292,	4·6526	378,219,	
'11	86,487,	4·6526	402,389,	44,025 home consumption
'21	91,753,	4·6526	426,890,	1,200 exported
'31	97,085,	4·6526	451,698, ⎬	—————
'41	102,452,	4·6526	476,668,	45,225 total in century
'51	107,852,	4·6526	501,792,	—————
'61	113,277,	4·6526	527,033,	
'71	118,703,	4·6526	552,278, ⎭	
2081	124,128,	4·6526	577,518, ⎫	
'91	129,527,	4·6526	602,637,	
2101	134,903,	4·6526	627,650,	68,803 home consumption
'11	140,231,	4·6526	652,439,	1,200 exported
'21	145,517,	4·6526	677,032,	—————
'31	150,742,	4·6526	701,342, ⎬	70,003 total in century
'41	155,897,	4·6526	725,326,	—————
'51	160,979,	4·6526	748,971,	
'61	165,986,	4·6526	772,266,	
'71	170,899,	4·6526	795,125, ⎭	
2181	175,719,	4·6526	817,550, ⎰⎱	8,176 home consumption 100 exported ————— 8,276 total for ten years
	Total consumption in 310 years			146,354

Note.—Total available coal in the United Kingdom, as estimated, 146,480 million of tons.

TABLE L. —*Showing the Estimated Population of Great Britain during the next Three Hundred and Ten Years from* 1871. *The Rate of Increase in* 1871, *viz.,* 12·533, *constantly Decreasing at the Rate of* 4·694 *per Cent. per Decade.*

GREAT BRITAIN.

Decade.	Increase per Cent. per Decade.	Estimated Future Population.
	Per cnt.	No.
1871...............	12·533	26,248,723
1881.........	11·94	29,383,000
'91...............	11·38	32,727,000
1901.............. .	10·85	36,278,000
'11...............	10·34	40,029,000
'21...............	9·86	43,975,000
'31...............	9·39	48,105,000
'41...............	8·95	52,411,000
'51...............	8·53	56,881,000
'61...............	8·13	61,505,000
'71...............	7·75	66,272,000
'81...............	7·39	71,170,000
'91...............	7·04	76,180,000
2001...............	6·71	81,292,000
'11...............	6·39	86,487,000
'21...............	6·09	91,753,000
'31...............	5·81	97,085,000
'41...............	5·53	102,452,000
'51...............	5·27	107,852,000
'61...............	5·03	113,277,000
'71...............	4·79	118,703,000
'81...............	4·57	124,128,000
'91...............	4·35	129,527,000
2101...............	4·15	134,903,000
'11...............	3·95	140,231,000
'21...............	3·77	145,517,000
'31...............	3·59	150,742,000
'41...............	3·42	155,897,000
'51...............	3·26	160,979,000
'61...............	3·11	165,986,000
'71...............	2·96	170,899,000
'81...............	2·82	175,719,000

TABLE M.—*Showing the Estimated Population of England and Wales during the next Three Hundred and Ten Years from 1871. The Rate of Increase in 1871, viz.,* 12·990, *constantly Decreasing at the Rate of* 4·563 *per Cent. per Decade.*

ENGLAND AND WALES.

Decade.	Increase per Cent. per Decade.	Estimated Future Population.
	Per cnt.	No.
1871...............	12·990	22,856,164
1881...............	12·40	25,690,000
'91...............	11·83	28,729,000
1901...............	11·29	31,973,000
'11...............	10·78	35,419,000
'21...............	10·28	39,061,000
'31...............	9·82	42,896,000
'41...............	9·37	46,916,000
'51...............	8·94	51,110,000
'61...............	8·53	55,470,000
'71...............	8·14	59,985,000
'81...............	7·77	64,646,000
'91...............	7·42	69,443,000
2001...............	7·08	74,358,000
'11...............	6·76	79,385,000
'21...............	6·45	84,506,000
'31...............	6·15	89,704,000
'41...............	5·87	94,969,000
'51...............	5·60	100,286,000
'61...............	5·35	105,652,000
'71...............	5·10	111,040,000
'81...............	4·87	116,448,000
'91...............	4·65	121,863,000
2101...............	4·44	127,274,000
'11...............	4·23	132,657,000
'21...............	4·04	138,016,000
'31...............	3·86	143,344,000
'41...............	3·68	148,620,000
'51...............	3·51	153,836,000
'61...............	3·35	158,990,000
'71...............	3·20	164,077,000
'81...............	3·05	169,081,000

TABLE N.—*Showing the Estimated Population of Scotland during the next Three Hundred and Ten Years from* 1871. *The Rate of Increase in* 1871, *viz.,* 9·550, *constantly Decreasing at the Rate of* 6·391 *per Cent. per Decade.*

SCOTLAND.

Decade.	Increase per Cent. per Decade.	Future Population per Decade.
	Per cnt.	No.
1871...............	9·550	3,392,559
1881...............	8·94	3,696,000
'91...............	8·37	4,005,000
1901	7·83	4,319,000
'11...............	7·33	4,635,000
'21...............	6·86	4,953,000
'31...............	6·43	5,272,000
'41...............	6·01	5,589,000
'51...............	5·63	5,903,000
'61...............	5·27	6,214,000
'71...............	4·93	6,521,000
'81...............	4·62	6,822,000
'91...............	4·32	7,117,000
2001...............	4·05	7,405,000
'11...............	3·79	7,686,000
'21...............	3·55	7,959,000
'31...............	3·32	8,223,000
'41...............	3·11	8,479,000
'51...............	2·91	8,725,000
'61...............	2·72	8,963,000
'71...............	2·55	9,191,000
'81...............	2·39	9,411,000
'91...............	2·23	9,621,000
2101...............	2·09	9,822,000
'11...............	1·96	10,014,000
'21...............	1·83	10,198,000
'31...............	1·71	10,372,000
'41...............	1·61	10,539,000
'51...............	1·50	10,697,000
'61...............	1·41	10,848,000
'71...............	1·32	10,991,000
'81...............	1·23	11,126,000

TABLE O.—*Estimate of the Prospective Increase of the Population of London during Three Hundred and Ten Years from 1871. Initial Rate 16·06 per Cent. per Decade. Decrement on ditto 13·046 per Cent.*

LONDON.

Decade.	Increase per Cent. per Decade.	Future Population per Decade.
	Per cnt.	No.
1871	16·06	3,254,260
1881	13·965	3,708,600
'91	12·144	4,158,800
1901	10·560	4,598,000
'11	9·182	5,020,200
'21	7·984	5,241,000
'31	6·943	5,797,400
'41	6·037	6,147,400
'51	5·250	6,470,100
'61	4·565	6,765,500
'71	3·970	7,034,100
'81	3·452	7,277,000
'91	3·001	7,495,300
2001	2·610	7,690,900
'11	2·270	7,865,400
'21	1·974	8,020,700
'31	1·716	8,158,300
'41	1·492	8,280,000
'51	1·298	8,387,500
'61	1·128	8,482,000
'71	0·981	8,565,200
'81	0·853	8,638,300
'91	0·742	8,702,400
2101	0·645	8,758,500
'11	0·561	8,807,600
'21	0·488	8,850,500
'31	0·424	8,888,100
'41	0·369	8,920,900
'51	0·321	8,949,500
'61	0·279	8,974,500
'71	0·242	8,996,200
'81	0·211	9,015,300

Plate 1.

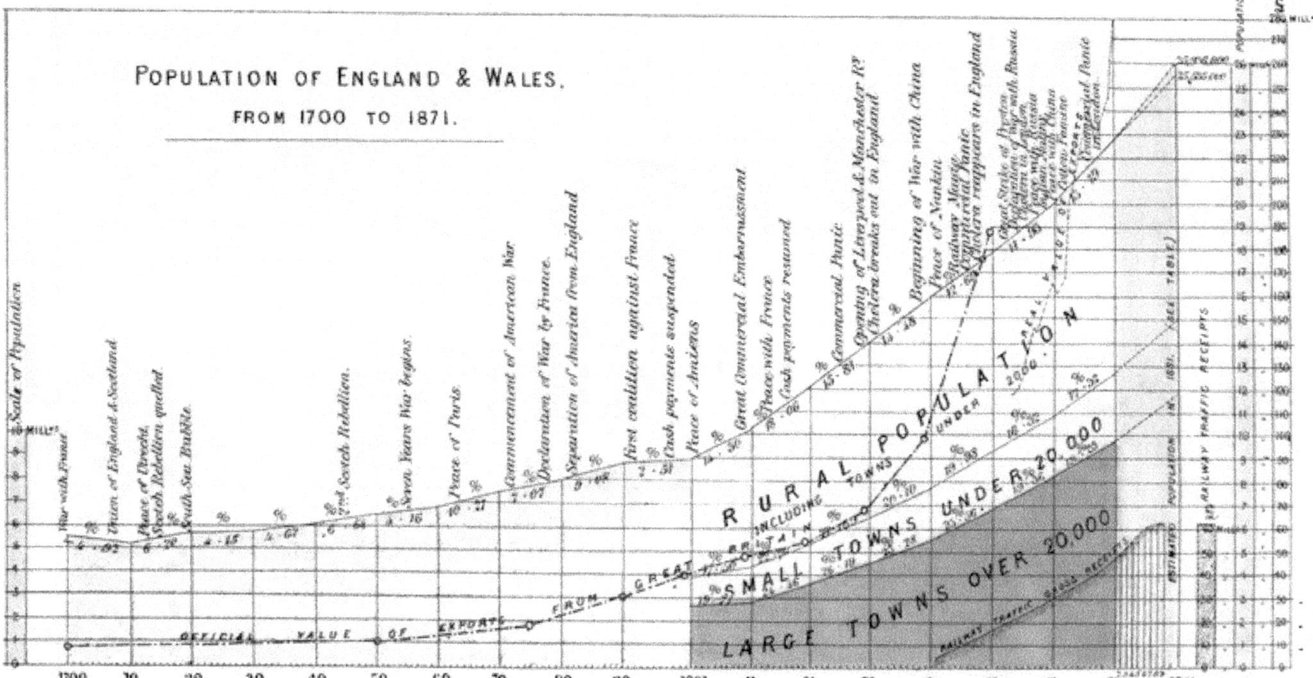

POPULATION OF ENGLAND & WALES.

FROM 1700 TO 1871.

Journal of Statistical Society, 1880.

NOTE. The figures immediately above the population of the small Towns give the decimal rates of increase of the total Town population; the uppermost figures those of the Total Population of England & Wales.

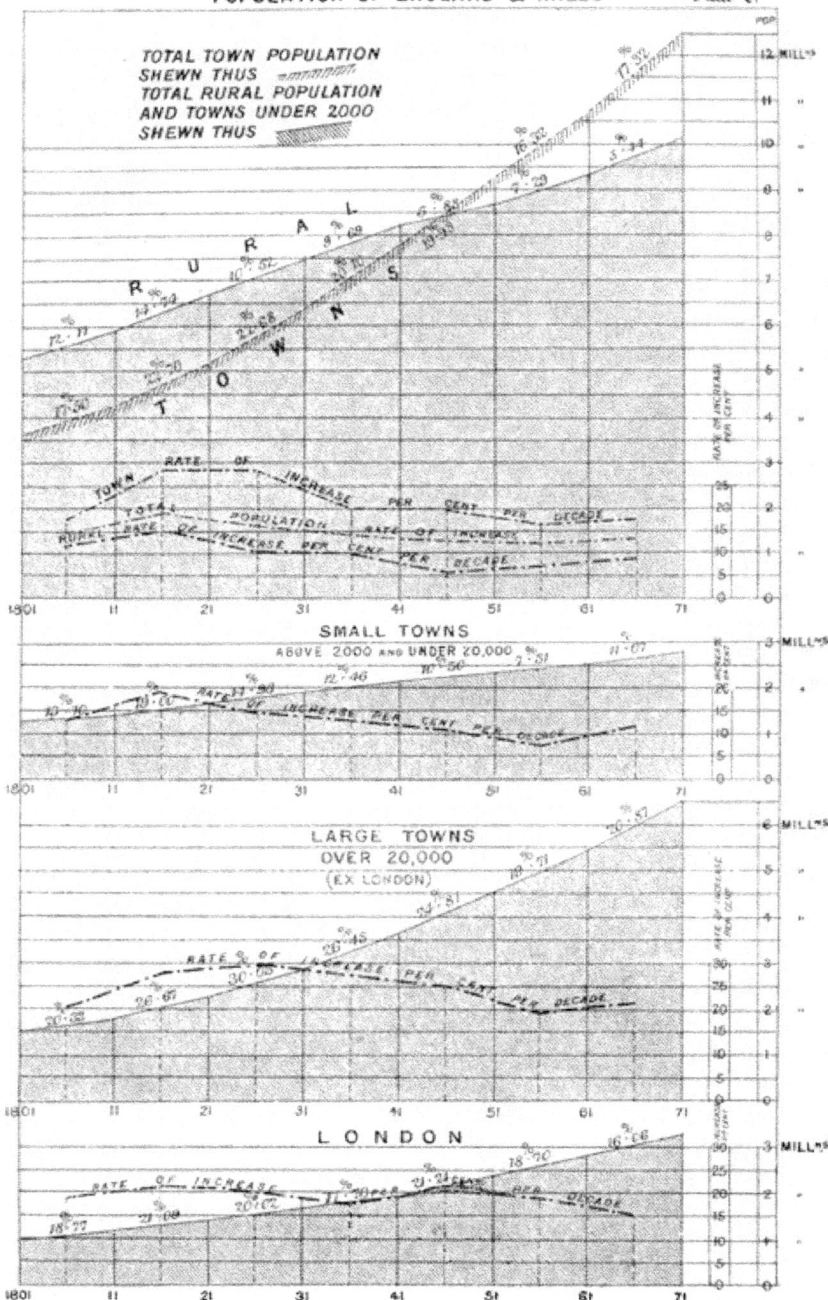

TOTAL TOWN POPULATION
SHEWN THUS

TOTAL RURAL POPULATION
AND TOWNS UNDER 2000
SHEWN THUS

R U R A L

T O W N S

TOWN RATE OF INCREASE PER CENT PER DECADE

TOTAL POPULATION

RURAL RATE OF INCREASE PER CENT PER DECADE

SMALL TOWNS
ABOVE 2000 AND UNDER 20,000

RATE OF INCREASE PER CENT PER DECADE

LARGE TOWNS
OVER 20,000
(EX LONDON)

RATE OF INCREASE PER CENT PER DECADE

L O N D O N

RATE OF INCREASE PER DECADE

POPULATION OF ENGLAND & WALES

Plate 3.

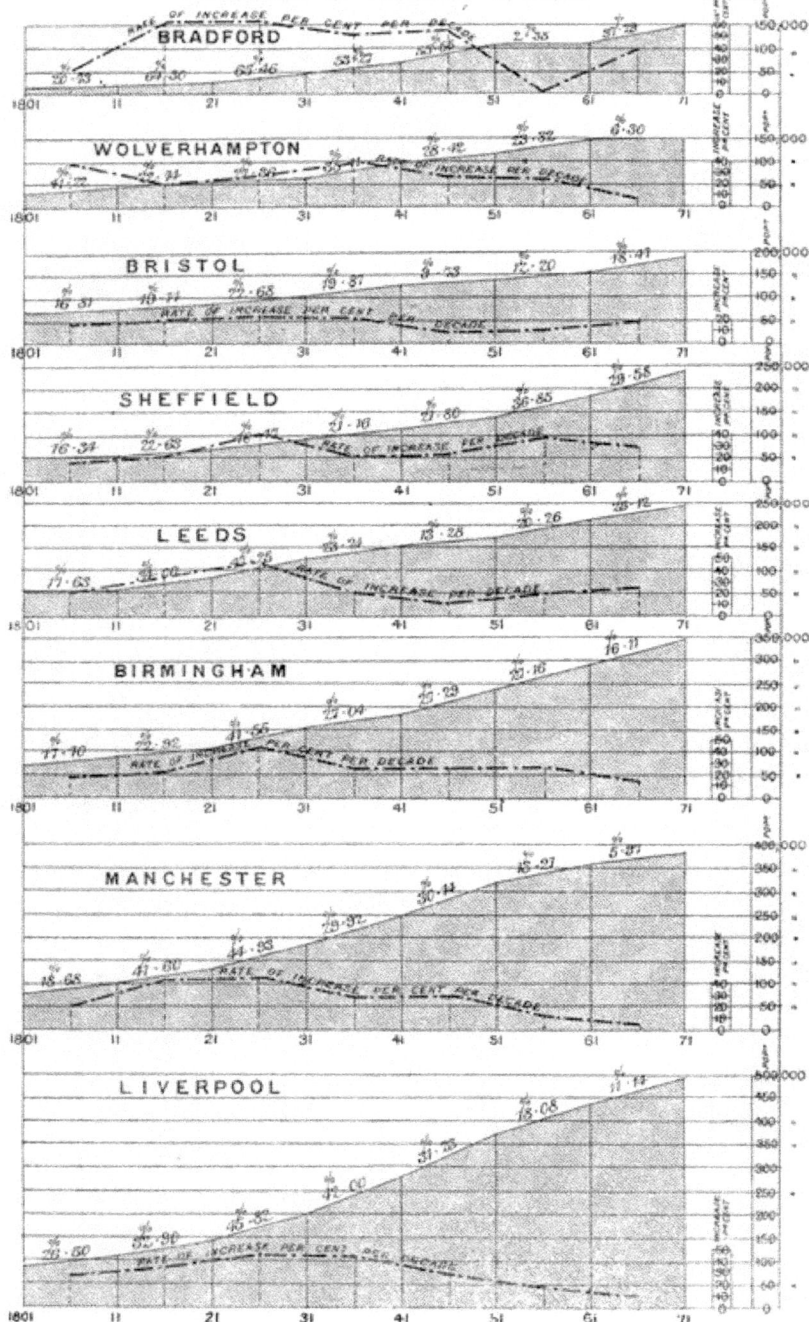

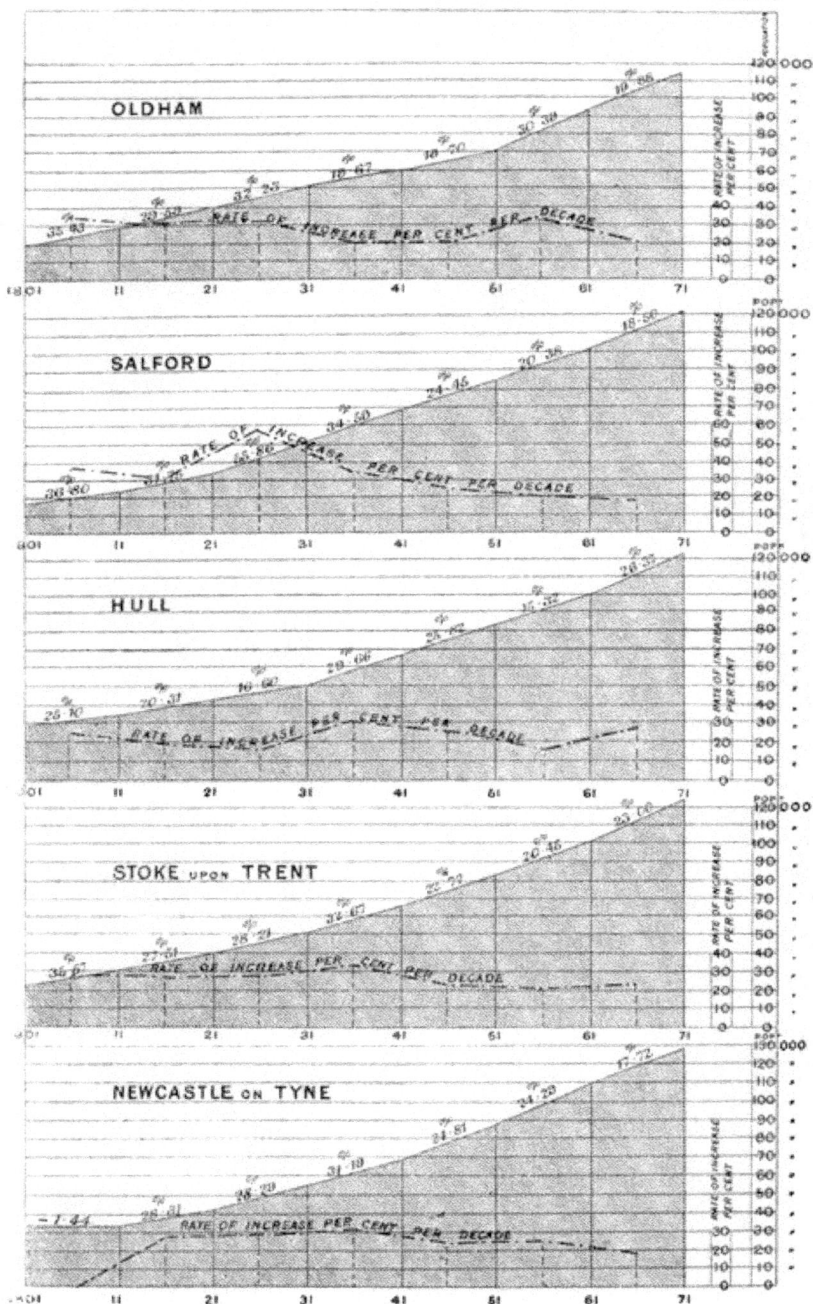

POPULATION OF ENGLAND & WALES. *Plate 4.*

OLDHAM

SALFORD

HULL

STOKE upon TRENT

NEWCASTLE on TYNE

POPULATION OF ENGLAND & WALES.

Plate 5.

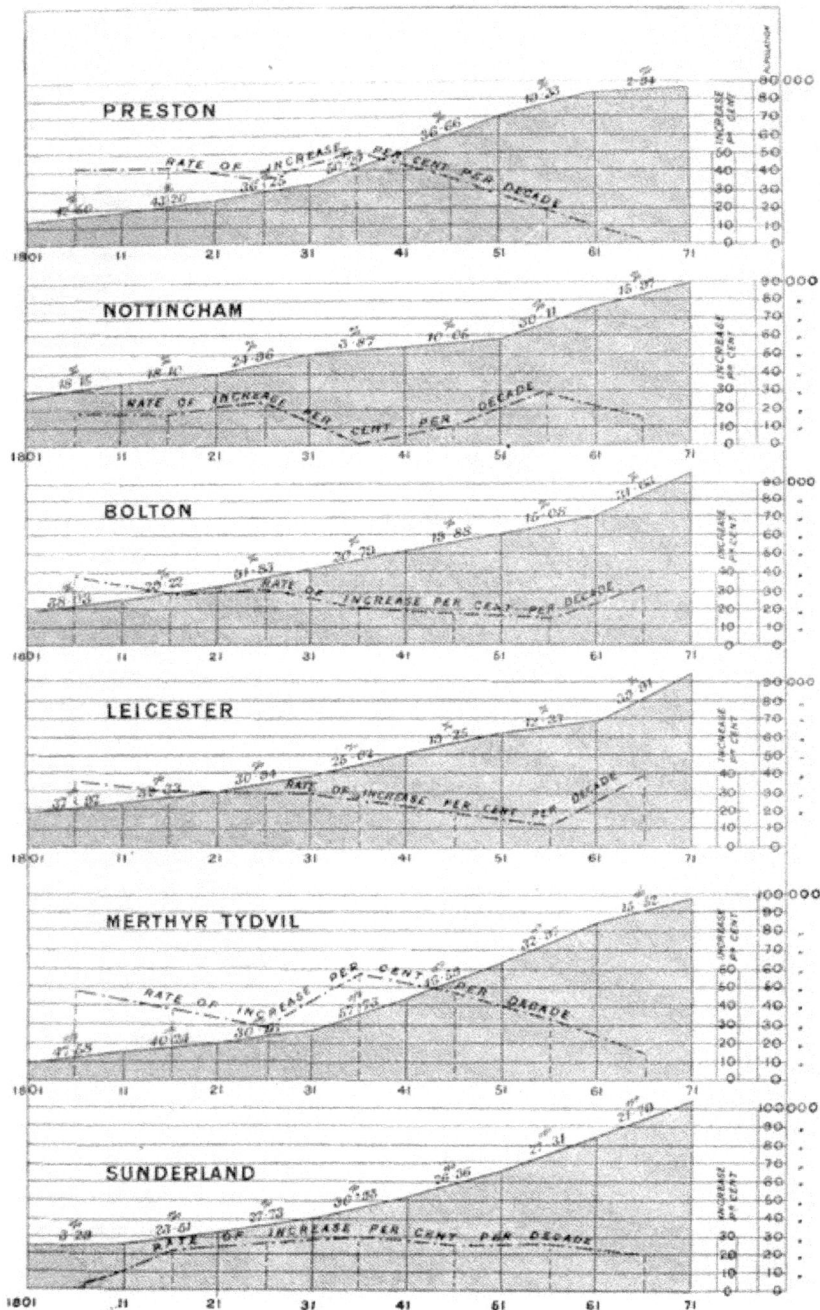

PRESTON

RATE OF INCREASE PER CENT PER DECADE

NOTTINGHAM

RATE OF INCREASE PER CENT PER DECADE

BOLTON

RATE OF INCREASE PER CENT PER DECADE

LEICESTER

RATE OF INCREASE PER CENT PER DECADE

MERTHYR TYDVIL

RATE OF INCREASE PER CENT PER DECADE

SUNDERLAND

RATE OF INCREASE PER CENT PER DECADE

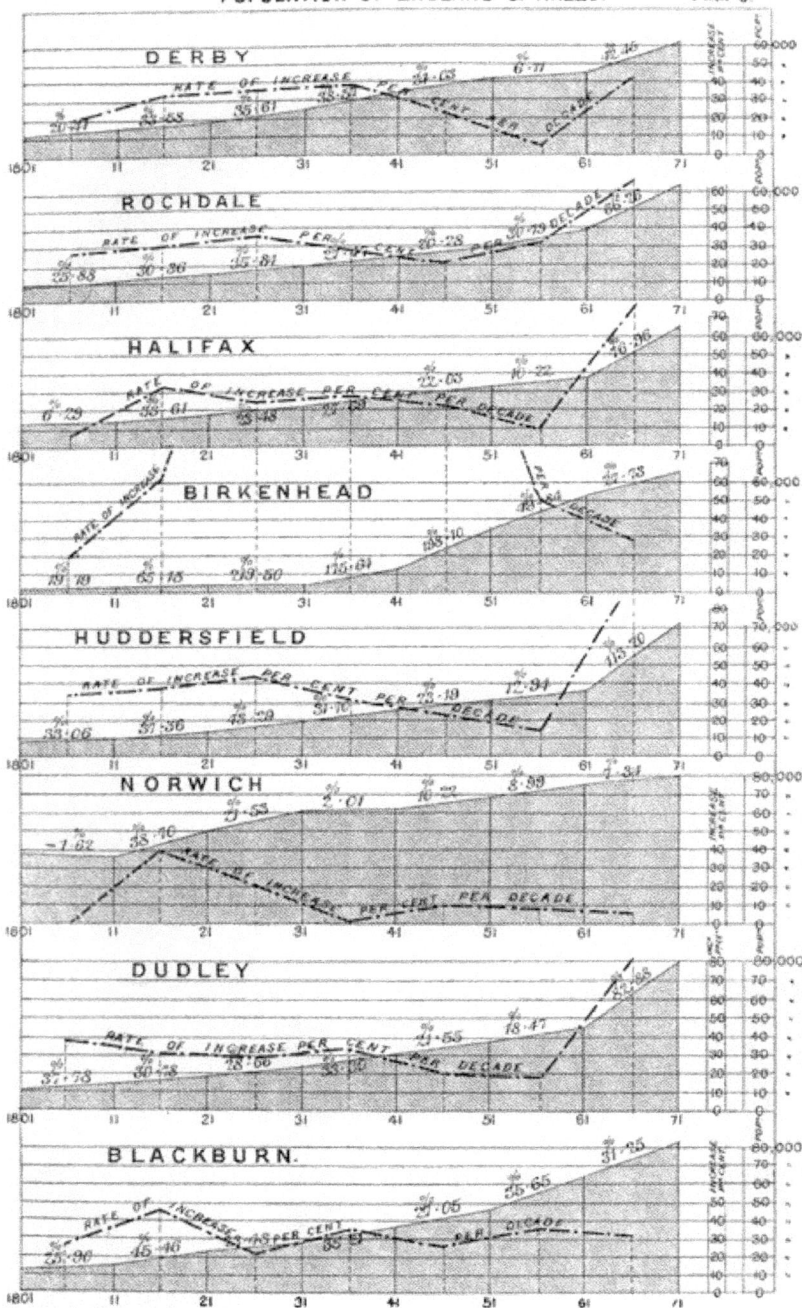

DERBY

RATE OF INCREASE PER CENT PER DECADE

20·21   33·88   35·81   28·20   24·63   6·71   12·26

1801   11   21   31   41   51   61   71

ROCHDALE

RATE OF INCREASE PER CENT PER DECADE

35·83   30·36   35·81   37   20·28   30·79   36·26

1801   11   21   31   41   51   61   71

HALIFAX

RATE OF INCREASE PER CENT PER DECADE

6·29   36·61   36·48   22·03   10·22   40·06

1801   11   21   31   41   51   61   71

BIRKENHEAD

RATE OF INCREASE PER CENT PER DECADE

29·89   65·18   249·50   115·04   133·10   43·84   72·73

1801   11   21   31   41   51   61   71

HUDDERSFIELD

RATE OF INCREASE PER CENT PER DECADE

33·06   37·36   43·39   31·70   73·19   12·94   113·70

1801   11   21   31   41   51   61   71

NORWICH

RATE OF INCREASE PER CENT PER DECADE

-1·82   18·70   5·55   2·01   10·   8·89   1·34

1801   11   21   31   41   51   61   71

DUDLEY

RATE OF INCREASE PER CENT PER DECADE

51·73   30·28   18·06   33   21·55   18·47   62·68

1801   11   21   31   41   51   61   71

BLACKBURN.

RATE OF INCREASE PER CENT PER DECADE

35·90   45·46   33   35·81   5·05   55·65   31·35

1801   11   21   31   41   51   61   71

Plate 7.

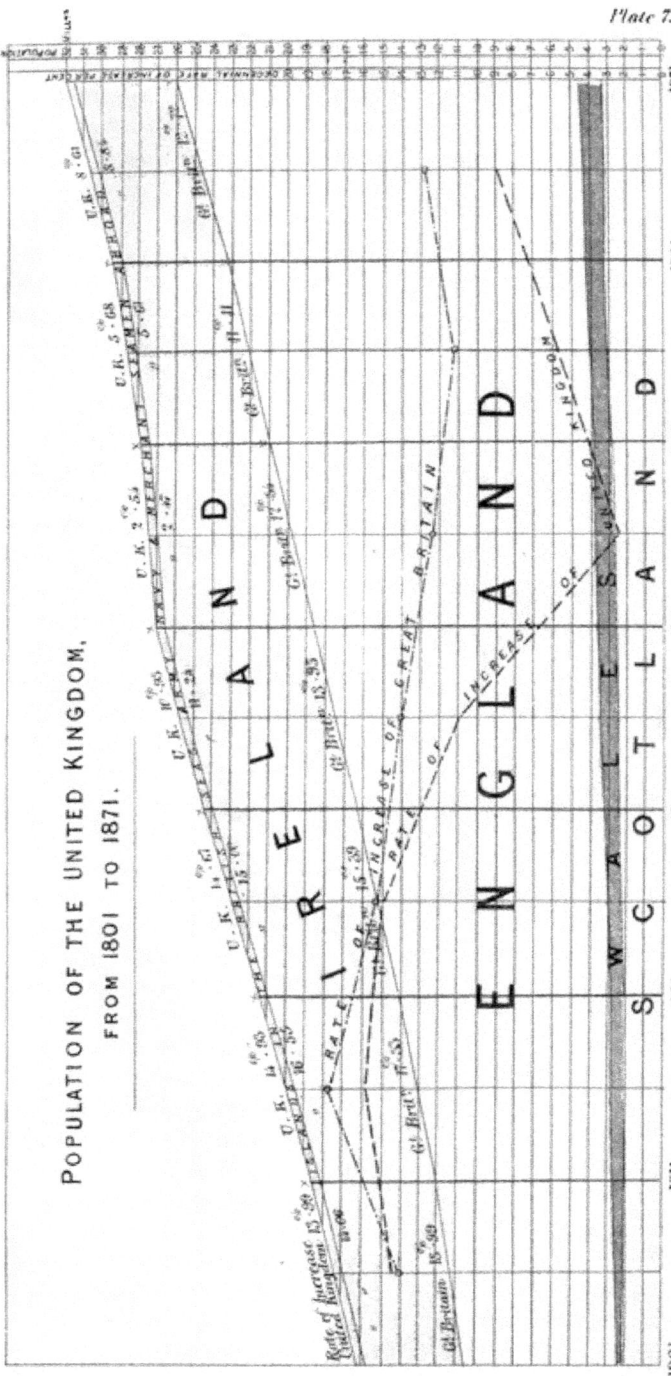

POPULATION OF THE UNITED KINGDOM,
FROM 1801 TO 1871.

ENGLAND

IRELAND

SCOTLAND

WALES

Diagram shewing the future estimated Population of Great Britain during 310 years from 1871, viz up to the year 2181, the computed date of exhaustion of Coal according to Table K.ᵃ

NOTE. The dotted line shews the Population Curve in the Diagram attached to the Report of the Coal Commission of 1871, the computed date of the exhaustion of Coal resulting from this Curve being A.D. 2231. (see Table K.

Plate 9

Plate 9

POPULATION

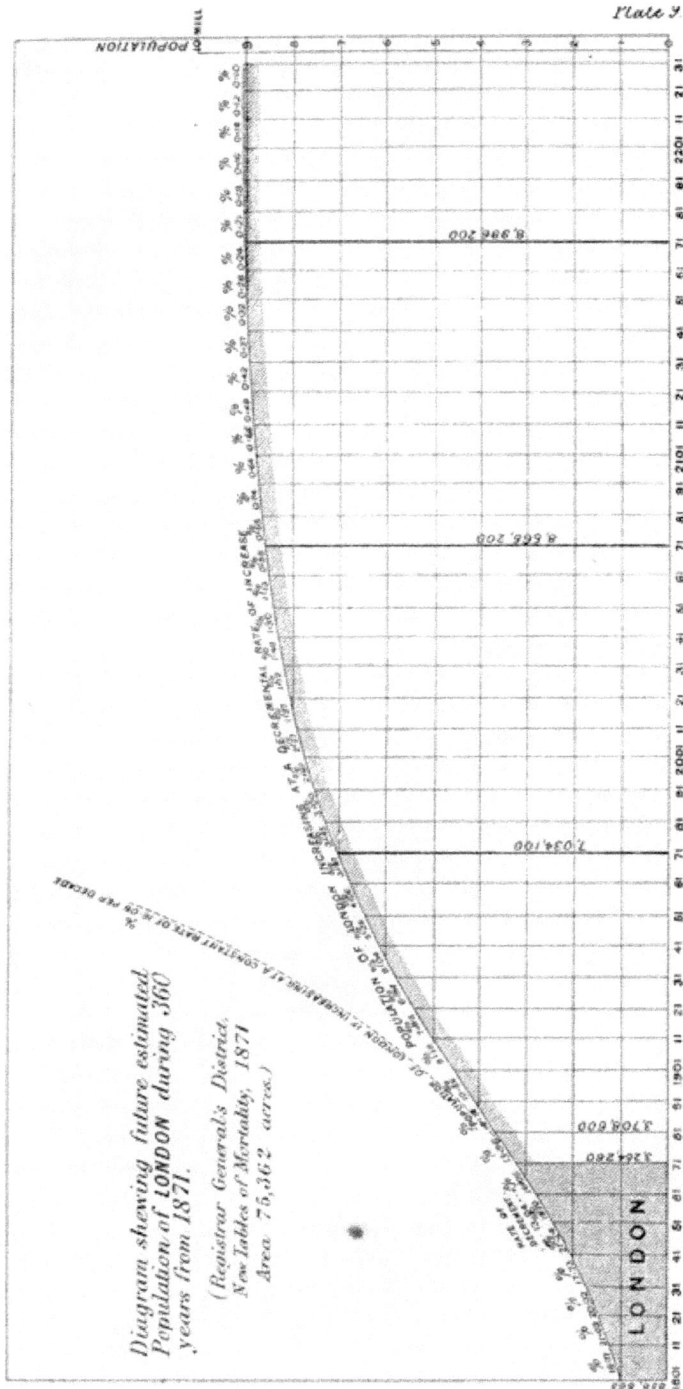

Diagram shewing future estimated
Population of LONDON during 360
years from 1871.

(Registrar General's District,
New Tables of Mortality, 1871
Area 75,362 acres.)

LONDON

8,396,200

8,666,200

7,034,100

3,708,600
3,254,260

DISCUSSION *on* MR. R. PRICE WILLIAMS'S PAPER.

SIR R. W. RAWSON, C.B., K.C.M.G., said he had devoted some time since yesterday morning to the study of the paper, and there were one or two points upon which he would say a few words; but before he entered into a consideration of the paper, he wished to say for himself, and he believed in doing so he would only give expression to the sentiments of every member of the Society, that he regretted very much that their esteemed colleague Dr. Farr (than whom no man in these islands, or perhaps in the world, would be better able to offer an opinion on this paper) was not present at that meeting. He rose thus early because he thought he could give a little interesting information on the subject, which he had derived from Mr. Price Williams's paper, and also because he wished to make a suggestion with regard to the validity of his deductions from the census returns. The paper was so valuable, and the materials collected had been obtained by such an amount of industry, that he was sorry he could not at once fully join with the author in his deductions, and admit the entire correctness of them. At the same time they were so important that he sincerely hoped that if Mr. Williams did not intend to publish the detailed statements from which his abstracts were printed, he would empower the Council of the Society to publish them, or to procure their publication by parliament or otherwise. Mr. Williams had put together the population of London in a series of years; that was very easy, but the separation of the population of all the other large towns, distinguished from the smaller towns, the aggregate of the smaller towns by themselves, and of the rural districts by themselves, was not light work, and the information thus supplied furnished such an amount of material for examination and for important deductions, that it ought not to be buried in manuscript. With regard to the doubt that occurred to him as to the validity of the author's deductions relating to the future increase of population, he found that between the years 1821 and 1861 there was a decrease, and between 1861 and 1871 an increase in the decennial increment. In 1821 the increase over 1811 was 18 per cent., the subsequent increase at each decennial period up to 1861 being $15\frac{3}{4}$, $14\frac{1}{2}$, $12\frac{3}{4}$, and nearly 12 per cent., from which the author argued that as in the fifty years from 1811 to 1861 there had been a gradual decrease in the increment, the same thing would go on till the end of his calculation—300 years. He (Sir R. W. Rawson) believed this arose not so much from a diminishing rate of increase in the population, as from an improvement in the enumeration of the population. At each census up to a recent period the increment caused by a more correct enumeration became successively smaller. The first census—that of 1801—was no doubt very imperfect, that of 1811 was somewhat less so. The experience of these two censuses enabled the commissioners in 1821 to make a very much better census, and he believed the decrease in the decennial increment up to a recent period arose to a great extent from the improved enumeration. This might not affect the whole of the calculations; but he believed it would have some influence upon them. He did not see why railways, manufactories, and steam engines—admitted to be causes of an increase of the population at an early period—should lose their influences, nor why, if it were so, the downward career should stop in 1861, and an increase in the opposite direction should exhibit itself in 1871. Then, of course, came in, as affecting the figures differently at different periods, the elements of the army and navy, the seafaring population, the channel islands, and so on; but he did not know how they affected the decrease; and without a scrutiny of the figures used by the experts of the Registration Office, he was not prepared to admit that they were to expect a constantly diminishing rate in the increment of the population. It was only natural that the increment should increase rateably if the country was prosperous; and he was not therefore prepared to admit that they were to calculate on a continuous decrease of 5 per cent. decennially in the rate of increment throughout the next 200 years. The other points which he wished to bring under their notice were much more gratifying to him, because they were not matters of doubt, but of certainty. Mr. Williams had brought together separately the populations of the large towns, of the small towns, and the rural districts at each census from 1801 down to 1871—a vast work for an individual—and one for which not only this Society, but every man in the kingdom who has occasion to look into these matters, must feel indebted to him, and he sincerely hoped the Council would endeavour to get the details published. If there were any elements of error in the abstracts contained in the paper, he hoped they would be eliminated, so that it might form the groundwork for future reference and comparison, and that it might be followed up by the commissioners of future censuses, so that the public might have the same classified abstracts in each census which Mr. Price Williams had made. He had drawn two deductions from the paper which he thought would be interesting to the meeting. The first was the relation of the four classes to one another under a common denomination, which the paper did not show. He had prepared a table which showed that. The second calculation which he had made from the paper was to show how far the augmentation of towns had arisen from the natural growth, namely, the excess of births over deaths, and how far from the influx of population, and how the rural districts had lost by the efflux, which had been drawn into the towns. He submitted these calculations to the members, and he hoped they would consider them of sufficient interest for him to be allowed to detain them a few minutes. The paper did not quite separate London from the other large towns, but he had done so. The only way in which he could ascertain the normal increase of population in England and Wales was by taking the average of the whole country, assuming the immigration and emigration of the whole to balance one another. Taking then the census of 1811 as compared with the preceding census of 1801 (always subject to the question of improved enumeration to which he had referred, and subject to any correction on account of a difference in the amounts of immigration and emigration during that decade, which he would suppose to be equal) for every 100 of increase on the average of the whole country, the increase in London was 138, in other large towns 140, in small towns 91 (the large towns having already begun to draw from the small ones to the extent of 9 in 100), and in the rural districts it was only 84; the migration from the rural districts into the towns then being 16 out of 100. He would run down each of these classes for the series of decades from 1811 to 1871, taking the average of the kingdom at 100. In London the increase was 138, 135, 165, 160; in 1851 it rose to 185, but in 1861 it fell to 162, and in 1871 was only 145; so that there was a gradual increase from 138 to 185 up to 1851, and a decrease from 185 in 1851 to 145 in 1871. It would be most interesting to add the figures of the approaching census, and to see what had happened during the last ten years, bringing the record up for seventy years—no small period in the life of a nation. For the large towns (exclusive of London) having over 20,000 inhabitants, the increase in the same decennial periods had been 140, commencing in 1811 with almost the same as in London (138), 134, 190, as compared with 165, 182, as compared with 160, and in 1851, 195, or nearly double the average of the kingdom, as compared with 185, which was also the maximum for London. In 1861 it fell from 195 to 165, and in 1871 to 152, London then being 145. With regard to the rest of the country, the excess was entirely on the side of the large towns and London, except in 1821, when there was a slight excess over the average in the small towns. In 1811 the increase in the small towns, instead of being 100, was only 91, then 105, 95, 86, 82, successively, and in 1861 it was only 61 (40 of their natural increase having gone to the augmentation of the large towns), while in 1871 it had risen to 83. But the most striking picture was in the rural districts. They began in 1811 with 84, then 81, 66, 67, down to 46 in 1851, when 54 per cent. of the normal increase had been drawn from the rural districts, rising afterwards to 61 and 64. These figures, he thought, would sufficiently clearly show how in London and the large towns the population had been drawn from the rural to the urban districts during the period under review. He had made some calculations with regard to the increment of each of these classes by natural growth, or excess of births over deaths, and by immigration or emigration. What he called the "natural growth" was of course affected by any difference in the rate of such growth in town and country districts respectively, and might be affected by an excess of immigration over emigration in the whole country, which would doubtless be directed towards the towns. During the decade from 1801 to 1811 the increase in London by natural growth was 137,000, and by immigration only 42,000. Then in each of the next three decades, the increase was 205,000, 218,000, and 239,000 by natural growth, and by immigration, 34,000, 58,000, and 53,000; but in 1851 came this change, the increase by natural growth was 247,000, and by immigration 166,000, or three times the amount of immigration of the preceding decade. In 1861 the increase by natural growth was 281,000 and by immigration 159,000. The effect of this immigration into London in those two decades being to increase the increment by natural growth in 1871 to 369,000—a very large augmentation, whilst the immigration was only 78,000, or about half the number in the preceding decade. He would be happy to hand the calculations he had made to Mr. Price Williams and the Council, if they thought worth while to examine them. With regard to the relative increase of these four classes between 1801 and 1871, the population of London had increased 339 per cent., the population now being more than three times what it was in 1801, while the large towns had increased 452 per cent., or 4½ times, the small towns 231 per cent., while the rural districts had only increased 92 per cent. There were other points to which he would refer had he not occupied so much time, but he thought what he had shown would indicate the immense value of the paper read by Mr. Price Williams, and he would again express the hope that the materials from which the tables furnished were derived would be printed in detail.

Note to Sir R. W. Rawson's Remarks.

No. 1.—Statement showing the Population of England and Wales, and of London, Large Towns, Small Towns, and Rural Districts separately; with the Percentage Increase of each Class separately, and of each Class compared with the Average of England and Wales, in each Decade from 1801 to 1871.

Years.	Total: England and Wales.	London.	Other Large Towns: with Population over 20,000.	Small Towns: with Population over 2,000 and under 20,000.	Rural Districts: including Towns under 2,000.
1801	8,892,536	958,863	1,445,290	1,211,092	5,277,291
'11	10,164,256	1,138,815	1,739,224	1,369,757	5,916,460
'21	12,000,236	1,378,947	2,203,082	1,630,046	6,788,661
'31	13,896,797	1,654,994	2,865,061	1,874,112	7,502,630
'41	15,909,132	1,948,417	3,623,758	2,107,562	8,229,395
'51	17,927,609	2,362,236	4,522,765	2,328,941	8,713,667
'61	20,066,224	2,803,989	5,414,220	2,499,051	9,348,964
'71	22,712,266	3,251,913	6,546,627	2,775,739	10,137,987
Percentage Increase in each Decade.					
1811	14·30	18·77	20·03	13·10	12·11
'21	18·06	21·09	24·37	19·00	14·74
'31	15·81	20·02	30·00	14·98	10·52
'41	14·48	17·73	26·47	12·46	9·69
'51	12·69	21·24	24·81	10·50	5·88
'61	11·93	18·70	19·71	7·31	7·29
'71	13·19	15·97	20·99	11·07	8·44

	Percentage Increase of each Class in each Decade, compared with the Average of England and Wales.				
	London.	Other Large Towns.	Average of England and Wales.	Small Towns.	Rural Districts.
1811	1·38	1·40	1·00	0·91	0·84
'21	1·35	1·34	1·00	1·05	0·81
'31	1·65	1·90	1·00	0·95	0·66
'41	1·60	1·82	1·00	0·86	0·67
'51	1·85	1·95	1·00	0·82	0·46
'61	1·62	1·65	1·00	0·61	0·61
'71	1·45	1·52	1·00	0·83	0·64

No. 2.—Statement of the Actual, Percentage, and Proportionate Rate of Increase of the Population of London, the Large and Small Towns respectively, and the Rural Districts of England and Wales, by Natural Growth (excess of Births over Deaths), and by Migration separately, in each Decade from 1811 to 1871.

	London.					
Years.	Increase in each Decade.		Percentage Rate of Increase.		Relative Percentage Proportion of Increase.	
	By Natural Growth.	By Immigration.	By Natural Growth.	By Immigration.	By Natural Growth.	By Immigration.
1811	137,117	42,835	14·30	4·46	76·2	23·8
'21	205,669	34,463	18·06	3·02	85·4	14·6
'31	218,011	58,036	15·81	4·21	79·0	21·0
'41	239,643	53,780	14·48	3·25	81·7	18·3
'51	247,254	166,565	12·69	8·55	59·8	40·2
'61	281,814	159,939	11·93	6·75	63·8	36·2
'71	369,846	78,078	13·19	2·78	82·6	17·4
Average	—	—	—	—	74·0	26·0

	Other Large Towns.					
1811	206,676	87,268	14·30	6·00	70·3	29·7
'21	314,103	149,755	18·06	8·61	67·7	32·3
'31	348,307	313,672	15·81	14·23	52·6	47·4
'41	414,860	343,837	14·48	12·00	54·7	45·3
'51	457,355	439,152	12·69	12·12	51·2	48·8
'61	539,565	351,890	11·93	7·77	60·6	39·4
'71	714,135	418,272	13·19	7·72	63·2	36·8

	Small Towns.			
	Increase in each Decade.		Percentage Rate of Increase or Decrease.	
	By Natural Growth.	By Immigration.	By Natural Growth.	By Immigration.
1811	173,186	Loss 14,521	14·30	− 1·20
'21	247,379	Gain 12,910	18·06	+ 0·94
'31	257,710	Loss 13,644	15·81	− 0·83
'41	271,391	„ 37,941	14·48	− 2·02
'51	267,449	„ 46,070	12·69	− 2·18
'61	277,842	„ 107,732	11·93	− 4·63
'71	329,624	„ 52,936	13·19	− 2·11

	Rural Districts.					
	Natural Increase.	Actual Increase.	Loss.	Increase by Natural Growth.*	Loss by Emigration.	Proportion of Increase which Migrated.
1811 ...	754,652	639,169	115,483	14·30	2·19	15·30
'21 ...	1,068,512	872,201	196,311	18·06	3·32	18·37
'31 ...	1,073,287	713,969	359,318	15·81	5·29	33·48
'41 ...	1,086,380	726,765	359,615	14·48	4·79	30·31
'51 ...	1,044,310	484,272	560,038	12·69	6·80	53·62
'61 ...	1,039,540	635,297	404,243	11·93	4·64	38·88
'71 ...	1,241,128	789,023	452,105	13·19	4·83	36·82
Total ...	—	—	2,447,113	—	—	—

* Assumed to be that of the average of England and Wales.

No. 3.—Percentage Proportions of the Town and Rural Population of England and Wales in 1801 and 1871, and the Actual Percentage Increase of each Class during that Period.

	Percentage Population.		Actual Percentage Increase in 1871.
	1801.	1871.	
London	10·8 ⎫	14·4 ⎫	339
Other large towns	16·3 ⎬ 40·7	28·8 ⎬ 55·4	452
Small towns	13·6 ⎭	12·2 ⎭	231
Rural districts	59·3	44·6	92
Total	100·0	100·0	—

No. 4.—Statement showing what the Population of the Rural Districts of England and Wales would have been if they had Increased from 1801 to 1871 in the same Proportion as the Population of the whole of England and Wales.

Increase from	5,277,291 in 1801	
To	13,473,219 ,, '71	= 155 per cent.
Instead of	10,137,987	= 92 ,,
A difference of	3,335,232	= 63 ,,

Mr. A. H. BAILEY (President of the Institute of Actuaries) remarked that Mr. Price Williams said, "In the absence of any census returns, the amount of the population prior to 1801 can only be approximately arrived at;" but he did not think the author had put sufficient emphasis on the very small reliance that could be placed on the returns prior to 1801. The poll tax was never levied on the whole population, the hearth tax excluded all cottages, and all the calculations of the numbers of the population were arrived at by an enumeration of the houses, and an estimate of the men, women, and children who were supposed to occupy them; and these calculations were frequently materially affected by political views. Thus in the parish of All Saints', Northampton, Dr. Price, from observations on the registers of baptisms and burials, came to the conclusion that the population was stationary—the fact being that there was an unusual proportion of Baptists in the town and parish who were not calculated, and by this hypothesis of a stationary population was brought about ("Northampton Table") an exceedingly erroneous table, and one which had caused a great deal of confusion. He thought little notice should be taken of the last century, for he could not believe that the rate of increase of the population in the last decade of the last century was only $2\frac{1}{2}$ per cent., while in the first decade of this century it was 14·3 per cent., and that during what was throughout for the most part a period of war; but he could not agree with Sir Rawson Rawson that little reliance was to be placed on the census returns in the early part of this century. No doubt improvements in taking the censuses had been made, but these were in collateral objects such as ages and occupations, rather than in counting the heads of the population. The early censuses in this respect, he thought, might reasonably be depended upon. There were two difficulties that presented themselves to his mind, one being that Scotland and Ireland were excluded from the returns. From both countries there was always, he thought, a considerable immigration into England; and the other was the influence of emigration to the colonies, the United States, and other parts of the world. He believed there were no trustworthy statistics of emigration until a comparatively recent date. Then as to the divisions made by Mr. Price Williams, there was a little ambiguity, because what was a rural population in 1801 was perhaps a town population in 1871. For instance, he found that in 1801 the population of Middlesborough-on-Tees was 239, whereas it was 46,621 in 1871, or an increase of about two hundred times, so that this would interfere with the general results. Then as to the question of boundaries. For instance, the question might be asked, what is London? There was the registration district of London, the southern division of which was, he thought, Streatham; but surely Croydon might be included in London. The extension of railways had brought people to be called Londoners who would not have been so called in the early part of the century, and he thought the question of boundary was a serious difficulty in all these calculations. He noticed that Brighton had made the greatest progress amongst the towns of 100,000 and upwards between 1801 and 1871.

Mr. N. A. HUMPHREYS thought they must all feel very much indebted to the writer of the paper for his wonderful industry, and for the immense amount of valuable facts which had been collected. With regard to the opinion expressed by the author, that the estimated increase of the population of the small towns, viz., 11·07 per cent., was too high, and that probably the actual population enumerated in 1881 would amount to no more than 25,500,000, he wanted to know what reason there was to support that opinion. They knew that up to 1861, there had been a steady decrease in the rate of increment, whereas an increase occurred between 1861 and 1871; and that all the facts since 1871 pointed to the most indubitable conclusion that the rate of increment of increase had been fully maintained since 1871. The annual natural increase of population was 11·9 per 1,000 in 1851-60, 12·6 in the decade 1861-70, and increased to 14·02 in the nine years ending 1879, showing that the actual increment has been more than maintained during the present decade. He thought the assumption, that in estimating the population we must take into account a continuance of an annual decrement in the increase was quite an erroneous one. Mr. Price Williams estimated the probable population at $25\frac{1}{2}$ millions in 1881, while Dr. Farr's method estimated it at 25,700,000. The registrar-general in 1871 estimated the population of London within 8,000 of the actual numbers returned; that was by taking into account the ascertained rate of increase during three preceding decades. In England and Wales, and in the manufacturing towns especially, it did not seem possible to invent a theory which would be applicable to a number of towns. The only satisfactory remedy for this difficulty being to hold a census more frequently than once in ten years.

Mr. CORNELIUS WALFORD thought the problems connected with the population of the last century had never had the attention paid to them which they deserved. There were special circumstances at work in the last century with which they were all familiar. The increase in the rate of population would always be about the same in each of the different great races. The Anglo-Saxon race had always been a prolific one, and it would in all probability remain so. In the last century there were continuous wars, and he would very much like to see (he had at one time intended to make one himself) an estimate of the drain of the population of this country by the wars, naval and military, for the number was so prodigious that he believed it would almost account for the want of increase in the population at that period. Besides the lives lost in the two services—army and navy—a very large migration to America took place by those who desired to escape compulsory service, as also from religious persecution. The armies serving in the field would not be a fair estimate, because foreigners (mercenaries) were frequently paid to fight for us. Regarding the subject of population generally, he had been weak enough, some years since, to read all the books on the subject that had been published in this country (he had about sixty of these in his own library), but he could not say that he knew much more about the actual facts after than he did before he began his reading, for the statements of the various authors were exceedingly conflicting; but there was one man whose writings commended themselves particularly to his judgment: he meant those of Mr. Rickman, whose report on the census of 1831 was well known. That was one of the very few books on population which the student might study with advantage. Mr. Bailey had anticipated him in referring to the question of the boundaries of some towns, especially with regard to Manchester and Liverpool, and perhaps there were no two towns in the country where the increase had been more marked or more continuous, and whatever statistics might say as to the apparent want of increase, those statistics were misleading, for the population spread entirely beyond the old boundaries. With regard to the increase of population between 1811 and 1821, all students of this question would know that after the drain of a great war nature seemed to reassert her sway, and production increases at more than the normal rate. He believed this would be found to have been the case in all countries. In 1851 there was another marked change, which he believed was due to the extension of railways and the adoption of free trade. Free trade largely developed our manufactures, and the towns grew very largely, and railways tended to take people into these large towns; but in later decades a reaction had set in; railways were no longer confined to the great centres; manufactories were being removed to smaller towns away from the great centres, in consequence of the economy of living in these smaller towns. He believed nearly every circumstance of prominence in the tables of the author could be explained and accounted for on rational principles, and there was not much fear but that the increase would go on in the future as in the past. It would, however, be a happy day for the country when, by the cultivation of habits of thrift, large families would no longer be associated with notions of pauperism and poor rates. He thanked the author for his paper.

Mr. FREDERICK HENDRIKS observed that it was interesting in inquiries relating to the progress of population, to look back upon some of the old forecasts of those who had published their speculations upon this subject, and to see how far they had been realised in the experience of the past. The estimates as to what might probably be the future population of England and Wales that were printed, in 1662, by Captain Graunt, a London citizen, under the auspices of the Royal Society, were based upon the most trustworthy and accurate methods of deduction applicable at that period. The leading data used by Captain Graunt were the numbers of deaths and of births recorded in the first half of the seventeenth century in certain London and country parishes, where the number of the inhabitants was known, perhaps roughly, but still with a sufficient approach to accuracy. It would appear that the population of England and Wales in 1662 was estimated at 6,440,000, or in round figures at $6\frac{1}{2}$ millions, of whom it was supposed that the people in and about London constituted a fifteenth part. Captain Graunt, after as careful a study as he could give to the London and country bills of mortality, came to the conclusion that the London mortality was about 1 in 32 over and above those who died of the plague, whilst the country mortality was not over 1 in 50 per annum. Now it is very curious to find that even if we throw into the estimate the mortality from the plague, and then take a total combined average of what may be called the general population mortality, urban and rural, as indicated by Captain Graunt's calculations in 1662, the result is an average of about $2\frac{1}{4}$ per cent. per annum, or say 1 in about $44\frac{1}{2}$. This does not differ to any material extent from the actual mortality of England and Wales now, in 1880, after the lapse of more than two centuries since the estimate was framed. This deserves to be considered by our sanguine sanitary reformers, who think that so vast a diminution has been taking place in the last two centuries in the rate of mortality. It also affords a certain degree of encouragement to such speculative inquiries as those now given to this Society in Mr. Price Williams's painstaking essay, and it is to be hoped that some statistician of two centuries hence will derive some such instruction from it as we of the present day can derive from looking back to old John Graunt's figures. As nearly as he (Mr. Hendriks) recollected what Captain Graunt said about the other factor in the growth of population, namely, the births, and their excess over deaths, it was made out that the result was to double the people in the country in about 280 years, and in London in about seventy years, the chief reason being that so many " breeders," as Captain Graunt called them, left the country for the metropolis, and bred there; whilst those who bred in the country were almost exclusively those who were born there. Of course at such rates of doubling of the population, the number of inhabitants in England and Wales at the present time would not be one-half of what it really is. Where then was the chief error in Captain Graunt's prognostications? Partly in its assuming that the number of children born to each family would not exceed four, whilst it has really been about five to six. This has greatly arisen also from his not foreseeing—although it is hard upon him to say he ought to have foreseen it—that the urban populations of this country, and the greater inducements and means for increase such populations would give rise to, as compared with those of rural populations, would receive a vast stimulus through this country becoming a manufacturing, instead of an essentially agricultural, community. Even in comparing the movements of population, such as Mr. Price Williams has given for various towns at decennial periods in this nineteenth century, we must cautiously consider whether we are really comparing like with like. In Lancashire and in several of the midland counties, for example, towns now closely peopled were, at some of these decennial periods, even within the memory of many present in this room, simply small country towns, or even only agglomerations of persons dependent chiefly on agriculture. These places are now the seats of flourishing manufactures and trades, affording such continuous employment to labour that higher ratios of marriage and of births to population have prevailed there than would have been experienced had they continuously remained what they formerly were. The exact growth of the change is in some cases more gradual than in others, but its effect on the increment of increase is obvious. The question is, after all, so large, and affected by so many distinctly disturbing causes, that it becomes of the first importance to study their leading elements in the estimate of the trustworthiness of any forecasts ahead, even for the next decennial period, and much more so in prognostications, interesting though they may be, which are to receive their fulfilment at so distant a date as is covered by Mr. Price Williams's figures.

Mr. G. HURST was of opinion that a different state of things would exist in the future as to the population of England and Wales. If the increase went on as at present the population would be trebled in a century. Our imports were increasing above the exports more and more, and he thought the checks enunciated by Malthus of prudence, poverty, and crime for restraining the population would come into operation very rapidly, and he did not believe that any real approximation could be arrived at with regard to the future population. We were depending so greatly on the foreigner for our food supplies that any disturbance in the supply might be a great check to the increase. If there come no very serious checks the country would be overrun with population, and it would be a great deal too crowded a century hence for anything but degradation and misery

Mr. S. BOURNE joined in the testimony of praise to Mr. Price Williams for his valuable paper, in which he thought the most

attractive feature was that which related to the future population ;
he could not but think that the rate of increase would be greater
and not smaller in the future. Half a century hence they might
look for quite double the present population of the United Kingdom, and his own opinion was that supposing no checks were
provided, they would be double the number in thirty-three and
a-half years. He believed that such a regulation of increase as that
to which Mr. Walford referred, was quite contrary to the dictates
of nature and prudence. What he would recommend was, that an
outlet should be found for the surplus population, and he could
hardly think it at all reasonable that we should be crowded here
whilst we have such magnificent countries in our own possession
only as yet partially developed. We have splendid food-producing
countries under our sway, and the remedy for over-population was
to facilitate the transfer of population to those places where every
increase of family is an increase of wealth. That he believed to be
the design of Providence—for we have had provided for us the
means of proper sustenance, enjoyment, happiness, and peace, if we
will only be wise enough to avail ourselves of them. The Americans
were more enlightened than we were, and were gradually going
from the seaboard to the far west, as the seaboard got too full, and
it should be our aim to develop Canada, Australia, and New
Zealand, not only because of reducing the population at home, but
because it would give us important countries upon which to rely for
our food supplies, as well as markets for our manufactures. This
was not only a question of sustenance, but of morals, happiness,
and peace, for people were herding here in barbarism, degradation,
and crime, when pure air, sunshine, and splendid regions were
waiting for them ; here they were a source of expense—there they
would be a source of wealth.

Mr. John B. Martin rose with diffidence to address a meeting
of experts upon the valuable paper which had been read, and the
supplementary paper by Sir Rawson Rawson. The Society could
appreciate the difficulties of dealing with the so-called official
returns of early periods, and could therefore sympathise with the
reader in being criticised on the one hand for basing deductions on
untrustworthy figures as regarded the early census returns, and
on the other for dismissing them in too summary a manner. He
would merely call attention to two points, first that every census
would probably have fewer numerical omissions than the preceding
one, and this would affect the apparent percentage of increase ; and
next, that while there was some attempt at recording the emigration from our seaports, no account was taken of the ebb and flow
of emigration and immigration across the channel, whereby our
chief commercial cities, and notably London, were filled with
foreigners, who displaced the native population, and very seriously
affected the apparent natural increment. This was a subject that
deserved careful observation.

Mr. Price Williams having referred to the diagrams exhibited,
replied to some points raised, and especially drew attention to the

remarkable effect of the continued decrease in the population of Ireland upon the rates of increase of the entire population of the United Kingdom.   As to the improvement in the census taken between 1811 and 1821 accounting for the very large amount of increase in that particular decade, he thought that assumption was discredited by the figures being put graphically on the diagram, as had there been an error due to an inaccuracy of that kind, it would be apparent at once on the diagram.   The cessation of war, the introduction of machinery, and the cheapening of food, no doubt had a great deal to do with the increment between 1841 and 1851, when all the great centres were opened up by means of railways. He had great diffidence in offering further observations as to his theory of a decremental rate, but he could not help thinking with Sir George Airy, that the theory would prove to be correct.   He thanked the meeting deeply and earnestly for their appreciation of his contribution on this important subject.

HARRISON AND SONS, PRINTERS IN ORDINARY TO HER MAJESTY, ST. MARTIN'S LANE.

In Monthly Parts, Price 3s.   Subscription 32s. per Annum.   Postage Free.

PRÉCIS

OF

OFFICIAL PAPERS,

BEING

ABSTRACTS OF ALL PARLIAMENTARY RETURNS

Directed to be Printed by both Houses of Parliament.

SESSION 1880.

LONDON: W. H. ALLEN AND CO.,

13, WATERLOO PLACE, S.W.,

Publishers to the India Office.

R. J. MITCHELL & SONS,

Booksellers, Bookbinders, &c.,

52 & 36, PARLIAMENT ST., & 52, BUCKINGHAM PALACE RD., S.W.

BOOKBINDING EXECUTED IN EVERY VARIETY OF STYLE, at *less* than Co-operative prices for Cash.   *Price List gratis.*

R. J. M. and Sons respectfully inform the Nobility and Gentry that every attention is paid to this particular branch, the best Workmen being employed.   The Sewing and Forwarding is strictly attended to, and a superior taste displayed in the Finishing.   At this Establishment a large assortment of Books of a superior character is kept constantly on sale, in various styles of Morocco and elegant Calf Bindings, from which specimens of Bookbinding may be selected as patterns for Binding.

Old Books Neatly Re-backed and Brightened Up at a Very Low Charge.

R. J. MITCHELL, LICENSED APPRAISER & VALUER, is prepared to value Libraries for Probate, &c., at a moderate Commission, also to purchase either small or large Collections of Books at a fair value.

DEPOT FOR THE SALE OF PARLIAMENTARY PAPERS,

AT GREATLY REDUCED PRICES.

36, PARLIAMENT STREET, S.W.

Libraries Purchased and Books Exchanged.

NOTED STORES FOR CHEAP MUSIC.

CO-OPERATIVE PRICES FOR CO-OPERATIVE TERMS. PROMPT CASH.

Note Address: **R. J. MITCHELL & SONS,**

52, BUCKINGHAM PALACE RD. (opposite the Grosvenor Hotel) LONDON, S.W.

STATISTICAL SOCIETY.

Honorary President.
HIS ROYAL HIGHNESS THE PRINCE OF WALES, K.G.

COUNCIL AND OFFICERS.—1880-81.

Honorary Vice-Presidents
(having filled the Office of President)

THE RIGHT HONOURABLE THE EARL OF SHAFTESBURY, K.G., D.C.L.

THE RIGHT HONOURABLE THE EARL OF HARROWBY, K.G., D.C.L.

THE RIGHT HONOURABLE THE LORD OVERSTONE, M.A., F.R.G.S.

THE RIGHT HONOURABLE THE EARL OF DERBY, D.C.L., F.R.S.

THE RIGHT HONOURABLE THE LORD HOUGHTON, D.C.L., F.R.S.

WILLIAM NEWMARCH, ESQ., F.R.S., F.I.A. (Corr. Member Inst. of France)

WILLIAM FARR, ESQ., M.D., C.B., D.C.L., F.R.S. (Corr. Member Inst. of France).

WILLIAM A. GUY, ESQ., M.B., F.R.C.P., F.R.S.

JAMES HEYWOOD, ESQ., M.A., F.R.S., F.G.S., &c.

GEORGE SHAW LEFEVRE, ESQ., M.P.

THOMAS BRASSEY, ESQ., M.P.

President.
JAMES CAIRD, ESQ., C.B., F.R.S.

Vice-Presidents.

HYDE CLARKE.
FREDERICK HENDRIKS.

PROF. W. S. JEVONS, LL.D., F.R.S.
FREDERIC JOHN MOUAT, M.D.

Trustees.

JAMES HEYWOOD, ESQ., M.A., F.R.S. | SIR JOHN LUBBOCK, BART., M.P., F.R.S.
WILLIAM NEWMARCH, ESQ., F.R.S.

Treasurer.
RICHARD BIDDULPH MARTIN, M.P.

Council.

ARTHUR H. BAILEY, F.I.A.
T. GRAHAM BALFOUR, M.D., F.R.S.
A. E. BATEMAN.
G. PHILLIPS BEVAN, F.G.S.
STEPHEN BOURNE.
EDWARD WILLIAM BRABROOK, F.S.A.
SIR GEORGE CAMPBELL, K.C.S.I., M.P.
J. OLDFIELD CHADWICK, F.R.G.S.
HAMMOND CHUBB, B.A.
HYDE CLARKE.
LIONEL L. COHEN.
MAJOR PATRICK G. CRAIGIE.
JULAND DANVERS.
ROBERT GIFFEN.

FREDERICK HENDRIKS.
NOEL A. HUMPHREYS.
PROF. W. S. JEVONS, LL.D., F.R.S.
ROBERT LAWSON.
PROFESSOR LEONE LEVI, LL.D.
JOHN B. MARTIN, M.A.
RICHARD BIDDULPH MARTIN, M.P.
FREDERIC JOHN MOUAT, M.D.
FRANCIS G. P. NEISON.
ROBERT HOGARTH PATTERSON.
HENRY D. POCHIN.
FREDERICK PURDY.
SIR RAWSON W. RAWSON, C.B., K.C.M.G.
CORNELIUS WALFORD, F.I.A.
THOMAS A. WELTON.

Secretaries.

HAMMOND CHUBB, B.A. | ROBERT GIFFEN.
JOHN B. MARTIN.

Foreign Secretary.
FREDERIC J. MOUAT, M.D.

Editor of the Journal.
ROBERT GIFFEN.

Assistant Secretary.
JOSEPH WHITTALL.

Bankers.—MESSRS. DRUMMOND AND CO., CHARING CROSS, S.W., LONDON.

www.ingramcontent.com/pod-product-compliance
Lightning Source LLC
Chambersburg PA
CBHW021533270326
41930CB00008B/1229